# Ergebnisse der Mathematik und ihrer Grenzgebiete

Band 66

Albrecht Pietsch

# Nuclear Locally Convex Spaces

Translated from the Second German Edition
by William H. Ruckle

Springer-Verlag Berlin Heidelberg New York
1972

Albrecht Pietsch
Friedrich-Schiller-Universität, Jena
Mathematisch-Naturwissenschaftliche Fakultät

William H. Ruckle
Department of Mathematical Sciences
Clemson University

An edition of this book is published by the Akademie-Verlag, Berlin, for distribution in socialist countries

Translation of the German edition:
Nukleare lokalkonvexe Räume, 2. Auflage
Copyright 1969 by Akademie-Verlag GmbH, Berlin

AMS Subject Classifications (1970) 46 A 05, 46 M 05, 47 B 10, 41 A 45

ISBN 3-540-05644-0 Springer-Verlag Berlin Heidelberg New York
ISBN 0-387-05644-0 Springer-Verlag New York Heidelberg Berlin

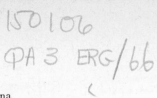

## Foreword to the First Edition

With a few exceptions the locally convex spaces encountered in analysis can be divided into two classes. First, there are the normed spaces, which belong to classical functional analysis, and whose theory can be considered essentially closed. The second class consists of the so-called nuclear locally convex spaces, which were introduced in 1951 by A. Grothendieck. The two classes have a trivial intersection, since it can be shown that only finite dimensional locally convex spaces are simultaneously normable and nuclear.

It can be asserted without exaggeration that the most important part of the theory of nuclear locally convex spaces is already contained in the fundamental dissertation of A. Grothendieck. Unfortunately, the machinery of locally convex tensor products which is used there is very cumbersome, so that many proofs are unnecessarily complicated. Thus, in this book I have undertaken the task of constructing the theory of nuclear locally convex spaces without using locally convex tensor products. Their place is taken by locally convex spaces of summable and absolutely summable families, which are introduced in the first chapter. The great advantage of this method is not merely the simplicity of the proofs. Indeed, the most important result is a necessary and sufficient condition for the strong topological dual of a nuclear locally convex space to be nuclear, which appears here for the first time.

In recent years certain concepts of approximation theory have attained great significance in the treatment of nuclear locally convex spaces. This statement is especially applicable to the concept of approximative dimension introduced by A. N. Kolmogorov and A. Pełczyński and applied by B. S. Mitiagin to characterize nuclear locally convex spaces. The principle results of these recent investigations are described in the three final chapters. We single out the Basis Theorem which has great significance for the representation theory of nuclear locally convex spaces. The main conclusion from these studies can be summarized in the following statement: If a theory of structure for locally convex spaces can be developed at all, then it must certainly be possible for nuclear locally convex spaces because these are more closely related to finite dimensional locally convex spaces than are normed spaces.

In order to present a clear narrative I have omitted exact references to the literature for individual propositions. However, each chapter begins with a short introduction which also contains historical remarks.

Deutsche Akademie der Wissenschaften zu Berlin
Institut für Reine Mathematik Albrecht Pietsch

# Foreword to the Second Edition

Since the appearance of the first edition, some important advances have taken place in the theory of nuclear locally convex spaces.

Firsts there is the Universality Theorem of T. and Y. Kōmura which fully confirms a conjecture of Grothendieck. Also, of particular interest are some new existence theorems for bases in special nuclear locally convex spaces. Recently many authors have dealt with nuclear spaces of functions and distributions. Moreover, further classes of operators have been found which take the place of nuclear or absolutely summing operators in the theory of nuclear locally convex spaces.

Unfortunately, there seem to be no new results on diametral or approximative dimension and isomorphism of nuclear locally convex spaces.

Since major changes have not been absolutely necessary I have restricted myself to minor additions. Only the tenth chapter has been substantially altered. Since the universality results no longer depend on the existence of a basis it was necessary to introduce an independent eleventh chapter on universal nuclear locally convex spaces. In the same chapter *s*-nuclear locally convex spaces are also briefly treated.

I have brought the bibliography up to date as well as I could. The newly added table of symbols ought to make the reader's job easier.

Friedrich-Schiller-Universität, Jena
Mathematisch-Naturwissenschaftliche Fakultät

Albrecht Pietsch

# Contents


**Chapter 0. Foundations** . . . . . . . . . . 1

0.1. Topological Spaces . . . . . . . . . . 1
0.2. Metric Spaces . . . . . . . . . . 3
0.3. Linear Spaces . . . . . . . . . . 4
0.4. Semi-Norms . . . . . . . . . . 5
0.5. Locally Convex Spaces . . . . . . . . . . 6
0.6. The Topological Dual of a Locally Convex Space . . . . . . . . . . 8
0.7. Special Locally Convex Spaces . . . . . . . . . . 10
0.8. Banach Spaces . . . . . . . . . . 11
0.9. Hilbert Spaces . . . . . . . . . . 11
0.10. Continuous Linear Mappings in Locally Convex Spaces . . . . . . . . . . 13
0.11. The Normed Spaces Associated with a Locally Convex Space . . . . . . . . . . 14
0.12. Radon Measures . . . . . . . . . . 16

**Chapter 1. Summable Families** . . . . . . . . . . 18

1.1. Summable Families of Numbers . . . . . . . . . . 18
1.2. Weakly Summable Families in Locally Convex Spaces . . . . . . . . . . 23
1.3. Summable Families in Locally Convex Spaces . . . . . . . . . . 25
1.4. Absolutely Summable Families in Locally Convex Spaces . . . . . . . . . . 27
1.5. Totally Summable Families in Locally Convex Spaces . . . . . . . . . . 29
1.6. Finite Dimensional Families in Locally Convex Spaces . . . . . . . . . . 32

**Chapter 2. Absolutely Summing Mappings** . . . . . . . . . . 34

2.1. Absolutely Summing Mappings in Locally Convex Spaces . . . . . . . . . . 34
2.2. Absolutely Summing Mappings in Normed Spaces . . . . . . . . . . 36
2.3. A Characterization of Absolutely Summing Mappings in Normed Spaces 38
2.4. A Special Absolutely Summing Mappings . . . . . . . . . . 42
2.5. Hilbert-Schmidt Mappings . . . . . . . . . . 45

**Chapter 3. Nuclear Mappings** . . . . . . . . . . 49

3.1. Nuclear Mappings in Normed Spaces . . . . . . . . . . 49
3.2. Quasinuclear Mappings in Normed Spaces . . . . . . . . . . 55
3.3. Products of Quasinuclear and Absolutely Summing Mappings in Normed Spaces . . . . . . . . . . 60
3.4. The Theorem of Dvoretzky and Rogers . . . . . . . . . . 67

**Chapter 4. Nuclear Locally Convex Spaces** . . . . . . . . . . 69

4.1. Definition of Nuclear Locally Convex Spaces . . . . . . . . . . 69
4.2. Summable Families in Nuclear Locally Convex Spaces . . . . . . . . . . 72
4.3. The Topological Dual of Nuclear Locally Convex Spaces . . . . . . . . . . 76
4.4. Properties of Nuclear Locally Convex Spaces . . . . . . . . . . 79

**Chapter 5. Permanence Properties of Nuclearity** . . . . . . . . . . . . . 85

5.1. Subspaces and Quotient Spaces . . . . . . . . . . . . . . . . . . . . 85
5.2. Topological Products and Sums . . . . . . . . . . . . . . . . . . . . 90
5.3. Complete Hulls . . . . . . . . . . . . . . . . . . . . . . . . . . . . 93
5.4. Locally Convex Tensor Products . . . . . . . . . . . . . . . . . . . 94
5.5. Spaces of Continuous Linear Mappings . . . . . . . . . . . . . . . . 95

**Chapter 6. Examples of Nuclear Locally Convex Spaces** . . . . . . . . . . 97

6.1. Sequence Spaces . . . . . . . . . . . . . . . . . . . . . . . . . . . 97
6.2. Spaces of Infinitely Differentiable Functions . . . . . . . . . . . . 99
6.3. Spaces of Harmonic Functions . . . . . . . . . . . . . . . . . . . . 102
6.4. Spaces of Analytic Functions . . . . . . . . . . . . . . . . . . . . 105

**Chapter 7. Locally Convex Tensor Products** . . . . . . . . . . . . . . . 107

7.1. Definition of Locally Convex Tensor Products . . . . . . . . . . . . 108
7.2. Special Locally Convex Tensor Products . . . . . . . . . . . . . . . 109
7.3. A Characterization of Nuclear Locally Convex Spaces . . . . . . . . 113
7.4. The Kernel Theorem . . . . . . . . . . . . . . . . . . . . . . . . . 115
7.5. The Complete $\pi$-Tensor Product of Normed Spaces . . . . . . . . . 117

**Chapter 8. Operators of Type $l^p$ and $s$** . . . . . . . . . . . . . . . 120

8.1. The Approximation Numbers of Continuous Linear Mappings in Normed Spaces . . . . . . . . . . . . . . . . . . . . . . . . . . . . . . 120
8.2. Mappings of Type $l^p$ . . . . . . . . . . . . . . . . . . . . . . . . 125
8.3. The Approximation Numbers of Compact Mappings in Hilbert Spaces . 129
8.4. Nuclear and Absolutely Summing Mappings . . . . . . . . . . . . . . 135
8.5. Mappings of Type $s$ . . . . . . . . . . . . . . . . . . . . . . . . . 138
8.6. A Characterization of Nuclear Locally Convex Spaces . . . . . . . . 141

**Chapter 9. Diametral and Approximative Dimension** . . . . . . . . . . . 144

9.1. The Diameter of Bounded Subsets in Normed Spaces . . . . . . . . . 144
9.2. The Diametral Dimension of Locally Convex Spaces . . . . . . . . . 149
9.3. The Diametral Dimension of Power Series Spaces . . . . . . . . . . 151
9.4. The Diametral Dimension of Nuclear Locally Convex Spaces . . . . 155
9.5. A Characterization of Dual Nuclear Locally Convex Spaces . . . . . 157
9.6. The $\varepsilon$-Entropy of Bounded Subsets in Normed Spaces . . . . . . . . 160
9.7. The Approximative Dimension of Locally Convex Spaces . . . . . . 164
9.8. The Approximative Dimension of Nuclear Locally Convex Spaces . . . 167

**Chapter 10. Nuclear Locally Convex Spaces with Basis** . . . . . . . . . . 171

10.1. Locally Convex Spaces with Basis . . . . . . . . . . . . . . . . . . 172
10.2. Representation of Nuclear Locally Convex Spaces with Basis . . . . . 173
10.3. Bases in Special Nuclear Locally Convex Spaces . . . . . . . . . . . 175

**Chapter 11. Universal Nuclear Locally Convex Spaces** . . . . . . . . . . 177

11.1. Imbedding in the Product Space $(\sum)^I$ . . . . . . . . . . . . . . . 177
11.2. Imbedding in the Product Space $(\sum')^I$ . . . . . . . . . . . . . . . 179

**Bibliography** . . . . . . . . . . . . . . . . . . . . . . . . . . . . . . . 181

**Index** . . . . . . . . . . . . . . . . . . . . . . . . . . . . . . . . . . 190

**Table of Symbols** . . . . . . . . . . . . . . . . . . . . . . . . . . . . 193

Chapter 0

# Foundations

This chapter presents the concepts and propositions which we shall use. Besides the most important definitions and a few elementary statements, all that we shall need from general topology is the fundamental Tichonov Theorem. The real foundation of our investigation is the theory of locally convex spaces. In addition to the Hahn Banach Theorem, we shall also need some other deep theorems. On the other hand, only some elementary statements on Hilbert space are necessary. This is also true for measure theory from which we take only the definition of Radon measure on a compact Hausdorff space and the construction of the Hilbert space $\mathfrak{L}^2_\mu(M)$.

References to these areas are:

Topology: N. Bourbaki [1], J. L. Kelley [1], G. Köthe [4].
Locally Convex Spaces: N. Bourbaki [4], A. Grothendieck [8], G. Köthe [4], J. L. Kelley and I. Namioka [1], H. H. Schaefer [1].
Normed Spaces: S. Banach [1], M. M. Day [1].
Hilbert Spaces: N. I. Ahiezer and I. H. Glazman [1], P. Halmos [2].
Measure Theory: N. Bourbaki [3], P. Halmos [1].

## 0.1. Topological Spaces

**0.1.1.** A **topology** $\mathfrak{T}$ is given on a set $M = \{x, y, \ldots\}$ if to each element $x \in M$ there correspondents a non-empty system $\mathfrak{U}_{\mathfrak{T}}(x)$ of subsets which satisfy the following conditions:

$(U_1)$ *If* $U \in \mathfrak{U}_{\mathfrak{T}}(x)$ *then* $x \in U$.

$(U_2)$ *For finitely many sets* $U_1, \ldots, U_n \in \mathfrak{U}_{\mathfrak{T}}(x)$ *there is a set* $U \in \mathfrak{U}_{\mathfrak{T}}(x)$ *with* $U \subset U_1 \cap \cdots \cap U_n$.

$(U_3)$ *Each set* $U \in \mathfrak{U}_{\mathfrak{T}}(x)$ *contains a subset* $U' \in \mathfrak{U}_{\mathfrak{T}}(x)$ *such that for each element* $y \in U'$ *there is a set* $V \in \mathfrak{U}_{\mathfrak{T}}(y)$ *with* $V \subset U$.

A set with a topology is called a **topological space.**

**0.1.2.** A subset $U$ of a topological space $M$ is called a **neighborhood of the element** $x$ if there is a set $U_0$ in $\mathfrak{U}_{\mathfrak{T}}(x)$ with $U_0 \subset U$. $\mathfrak{U}_{\mathfrak{T}}(x)$ is then referred to as a **fundamental system of neighborhoods of the element** $x$.

**0.1.3.** A topology $\mathfrak{T}_1$ is **finer** than a topology $\mathfrak{T}_2$ if each $\mathfrak{T}_2$-neighborhood of an arbitrary element $x \in M$ is also a $\mathfrak{T}_1$-neighborhood. The topology $\mathfrak{T}_2$ is then said to be **coarser** than the topology $\mathfrak{T}_1$. Two topologies on a set $M$ are **equal** if all elements $x \in M$ have the same neighborhoods.

**0.1.4.** A topology can be introduced on each subset $M'$ of a topological space $M$ by taking the collection of intersections $U' = M' \cap U$ with $U \in \mathfrak{U}_{\mathfrak{F}}(x)$ as a fundamental system of neighborhoods for each element $x \in M'$. This is called the **topology induced on** $M'$ by $M$.

**0.1.5.** A **directed system** $\{x_\alpha\}$ in a topological space $M$ is a set of elements $x_\alpha \in M$ which are uniquely associated with the elements of an index set $A$. For certain pairs of indices $\alpha$ and $\beta$ in $A$ a relation $\alpha \geqq \beta$ (read: $\alpha$ is greater than $\beta$) is defined which has the following properties:

($G_1$) *For* $\alpha, \beta, \gamma \in A$, $\alpha \geqq \beta$ *and* $\beta \geqq \gamma$ *always implies* $\alpha \geqq \gamma$.

($G_2$) *For finitely many indices* $\alpha_1, \ldots, \alpha_n \in A$ *there is an index* $\alpha \in A$ *with* $\alpha \geqq \alpha_1, \ldots, \alpha \geqq \alpha_n$.

Directed systems over the set of natural numbers are called **sequences**.

A directed system $\{x_\alpha\}$ **converges** to an element $x \in M$, if for each neighborhood $U \in \mathfrak{U}_{\mathfrak{F}}(x)$ there is an index $\alpha_0 \in A$ such that

$$x_\alpha \in U \qquad \text{for } \alpha \geqq \alpha_0 .$$

This is written

$$\lim_\alpha x_\alpha = x ,$$

and $x$ is designated as the **limit** of the directed system $\{x_\alpha\}$.

The limit of each convergent directed system in a topological space is uniquely determined if and only if the following condition is valid:

($U_4$) *For each pair of distinct elements* $x, y \in M$ *there are always sets* $U \in \mathfrak{U}_{\mathfrak{F}}(x)$ *and* $V \in \mathfrak{U}_{\mathfrak{F}}(y)$ *with* $U \cap V = \emptyset$.

A topological space satisfying ($U_4$) is called a **Hausdorff space.**

**0.1.6.** A subset $G$ of a topological space $M$ is **open** if for each element $x \in G$ there is a set $U \in \mathfrak{U}_{\mathfrak{F}}(x)$ with $U \subset G$. The empty set is considered open.

**0.1.7.** The **closed hull** $\overline{S}$ of an arbitrary subset $S$ of a topological space $M$ consists of all elements $x \in M$ for which every intersection $S \cap U$ with $U \in \mathfrak{U}_{\mathfrak{F}}(x)$ is non-empty. Each subset is contained in its closed hull. Those subsets which coincide with their closed hull are said to be **closed**. The empty set is considered closed as well as open.

**0.1.8.** A subset $D$ of a topological space $M$ is **dense** in $M$ if its closed hull $\overline{D}$ coincides with $M$. A topological space is **separable** if it contains a countable dense subset.

**0.1.9.** A Hausdorff space $M$ is **compact** if each collection of open sets which covers $M$ contains a finite subcollection which also covers $M$.

**Proposition.** *If $M$ is a compact Hausdorff topological space with the topology $\mathfrak{T}$ and $\mathfrak{T}'$ is a coarser Hausdorff topology on $M$, then $\mathfrak{T}$ and $\mathfrak{T}'$ must be equal.*

A subset $K$ of a topological space $M$ is **compact** if $K$ is a compact Hausdorff space with respect to the topology induced by $M$.

**0.1.10.** If $M_i$ with $i \in I$ is a collection of topological spaces then the set $M = \prod_I M_i$ of all families $x = [x_i, I]$ with $x_i \in M_i$ can also be made into a topological space by associating with each element $[x_i, I] \in M$ the fundamental system of neighborhoods which consists of all sets

$$U = [U_i, I] = \{[y_i, I] : y_i \in U_i\} .$$

Here the neighborhoods $U_i$ of the elements $x_i$ are so chosen that $U_i$ is equal to $M_i$ except for finitely many indices $i \in I$. The topological space $M$ thus obtained is called the **topological product** of the topological spaces $M_i$.

**Tichonov Theorem**. *The topological product of an arbitrary collection of compact Hausdorff spaces is also a compact Hausdorff space.*

**0.1.11.** Let $M$ and $M'$ be two topological spaces. A mapping $T$ from $M$ into $M'$ is called **continuous** if the pre-image

$$G = T^{-1}(G') = \{x \in M : Tx \in G'\}$$

of each open subset $G'$ of $M'$ is an open subset of $M$.

## 0.2. Metric Spaces

**0.2.1.** A mapping $\varrho$, which determines a non-negative number $\varrho(x, y)$ for each pair of elements $x$ and $y$ from a set $M$ is called a **metric** if it has the following properties:

($M_1$) *For $x, y \in M$ the statements $\varrho(x, y) = 0$ and $x = y$ are equavalent.*

($M_2$) *For $x, y \in M$, $\varrho(x, y) = \varrho(y, x)$.*

($M_3$) *For three elements $x, y, z \in M$ the triangle inequality*

$$\varrho(x, z) \leqq \varrho(x, y) + \varrho(y, z)$$

*is valid.*

**0.2.2.** A Hausdorff topology $\mathfrak{T}$ can be obtained for every set $M$ on which a metric $\varrho$ is defined by taking the collection of subsets

$$U_\varepsilon(x) = \{y \in M : \varrho(x, y) \leqq \varepsilon\} \quad \text{with } \varepsilon > 0$$

as a fundamental system of neighborhoods for each element $x \in M$. A topological space is called **metric** if its topology can be derived from a metric in this way.

**0.2.3.** Many proofs are easier in metric spaces because all discussion of convergence can be restricted to sequences. For instance, a subset $S$ is closed if and only if the limit of each convergent sequence in $S$ is also in $S$.

**0.2.4. Proposition.** *A metric space is compact if and only if each sequence of elements contains a convergent subsequence.*

**0.2.5.** A metric space $M$ is called **precompact** if for each positive number $\varepsilon$ there are finitely many elements $x_1, \ldots, x_s \in M$ such that

$$M = \bigcup_{n=1}^{s} U_\varepsilon(x_n) .$$

Every compact metric space is precompact.

**Proposition.** *Each precompact metric space is separable.*

## 0.3. Linear Spaces

**0.3.1.** A **linear space** $E$ over a commutative field $K = \{\alpha, \beta, \ldots\}$ is a set of elements $x, y, z, \ldots$ which has the following algebraic structure: For each two elements $x, y \in E$ a sum $x + y \in E$ is determined, and for each $\alpha \in K$ and all $x \in E$ a product $\alpha x \in E$ is defined. The following rules apply:

$x + y = y + x$ for $x, y \in E$. (1)

$x + (y + z) = (x + y) + z$ for $x, y, z \in E$. (2)

There is a zero element $o \in E$ with $x + o = x$ for $x \in E$. (3)

$1 x = x$ for $x \in E$.

$0 x = o$ for $x \in E$.

$\alpha(\beta x) = (\alpha \beta) x$ for $\alpha, \beta \in K$ and $x \in E$. (6)

$\alpha (x + y) = \alpha x + \alpha y$ for $\alpha \in K$ and $x, y \in E$. (7)

$(\alpha + \beta) x = \alpha x + \beta x$ for $\alpha, \beta \in K$ and $x \in E$. (8)

In the sequel $K$ is always the field of real or complex numbers. Accordingly $E$ will be called a **real** or **complex linear space**.

**0.3.2.** A sum $\alpha_1 x_1 + \cdots + \alpha_n x_n$ with $\alpha_1, \ldots, \alpha_n \in K$ is called a **linear combination** of the elements $x_1, \ldots, x_n \in E$. The elements $x_1, \ldots, x_n$ are said to be **linearly independent** if $\alpha_1 x_1 + \cdots + \alpha_n x_n = o$ always implies $\alpha_1 = \alpha_2 = \cdots = \alpha_n = 0$. A linear space $E$ is **n-dimen-**

**Mackey Theorem.** *Each weakly bounded subset of a locally convex space is bounded.*

**0.6.4.** For every subset $M$ of a locally convex space $E$ the (absolute) **polar**

$$M^0 = \{a \in E' : |\langle x, a\rangle| \leqq 1 \quad \text{for } x \in M\}$$

is a weakly closed and absolutely convex subset of $E'$. The **bipolar**, $M^{00}$, which contains $M$, is obtained by taking the polar of $M^0$. A more precise relation between $M$ and $M^{00}$ is given by the

**Bipolar Theorem.** *For each absolutely convex subset $M$ of a locally convex space, $M^{00}$ coincides with the closed hull of $M$.*

For each linear subspace $F$ of $E$, $F^0$ is a linear subspace of $E'$ and

$$F^0 = \{a \in E' : \langle x, a\rangle = 0 \quad \text{for } x \in F\} .$$

**0.6.5. Alaoğlu Bourbaki Theorem.** *For each zero neighborhood $U$ of a locally convex space $E$, the polar $U^0$ is a weakly compact subset of $E'$.*

**0.6.6.** A set $G$ of linear forms which are defined on a locally convex space $E$ is called **equicontinuous** if there is a neighborhood $U \in \mathfrak{U}(E)$ with $G \subset U^0$.

**0.6.7.** The collection of semi-norms

$$p'_K(a) = \sup \{|\langle x, a\rangle| : x \in K\} ,$$

where $K$ ranges over all precompact subsets of a locally convex space $E$, makes the topological dual $E'$ into a locally convex space $E'_k$. The following statement is valid:

**Proposition.** *On each equicontinuous subset of the topological dual of a locally convex space $E$, the topologies induced by $E'_s$ and $E'_k$ coincide.*

**0.6.8.** The so-called **strong topology** on the topological dual of a locally convex space is obtained from the system of semi-norms

$$p'_B(a) = \sup \{|\langle x, a\rangle| : x \in B\} \quad \text{with } B \in \mathfrak{B}(E) .$$

The locally convex space thus obtained is denoted by $E'_b$.

From the equation

$$B^0 = \{a \in E' : p'_B(a) \leqq 1\}$$

we have

$$p'_B(a) = \inf \{\varrho > 0 : a \in \varrho\, B^0\} .$$

Because of this relation it is easy to see that the polar sets $B^0$ with $B \in \mathfrak{B}(E)$ form a fundamental system of zero neighborhoods for the strong topology. This remains true when $B$ ranges over a fundamental system $\mathfrak{B}_{\mathfrak{F}}(E)$ of bounded subsets.

All statements and concepts with respect to the strong topology will be distinguished by the word **"strongly"**.

**0.6.9.** The topological dual of $E'_b$ is designated as the topological **bidual** $E''$ of the locally convex space $E$. Since $\langle x, a\rangle$ for fixed $x$ is a continuous linear form on $E'_b$, $E$ can be identified with a linear subspace of $E''$.

The so-called **natural topology** on $E''$ is obtained from the semi-norms

$$p'_{U^0}(x) = \sup\{|\langle x, a\rangle| : a \in U^0\} \quad \text{with } U \in \mathfrak{U}(E)$$

defined for $x \in E''$. The locally convex space thus obtained is denoted by $E''_n$. The natural topology induced on $E$ by $E''_n$ coincides with the original topology of $E$.

## 0.7. Special Locally Convex Spaces

**0.7.1.** A locally convex space $E$ is called **quasi-barrelled** if each strongly bounded subsets of $E'$ is equicontinuous. This means that on the topological bidual $E''$ the strong topology is the same as the natural topology so that $E$ can be considered as a linear subspace of $E''_b$. It is easy to see that quasi-barrelled locally convex spaces $E$ can be characterized by the following property: A closed absolutely convex subset $U$ is a zero neighborhood if for each bounded set $A$ there is a positive number $\varrho$ with $A \subset \varrho\, U$.

If all countable strongly bounded subsets of the topological dual of a locally convex space $E$ are equicontinuous, then we say that $E$ is **$\sigma$-quasi-barelled**.

**0.7.2.** A locally convex space $E$ is called **semi-reflexive** if $E$ coincides with $E''$. This happens if and only if every set from $\mathfrak{B}(E)$ is weakly compact.

**0.7.3.** A locally convex space $E$ is called **reflexive** if it is both quasi-barrelled and semi-reflexive. Then $E$ can be identified topologically as well as algebraically with its bidual $E''_b$ and there is a complete symmetry between $E$ and $E'_b$ because each of the two locally convex spaces is the strong topological dual of the other.

**0.7.4.** If a locally convex space $E$ has a countable fundamental system of zero neighborhoods it is called **metrizable** because its topology can be derived from a metric. A complete metrizable locally convex space is called an $(F)$-**space**.

**Proposition.** *Each metric locally convex space is quasi-barrelled.*

**0.7.5.** A $\sigma$-quasi-barrelled locally convex space $E$ in which a countable fundamental system of bounded subsets exists will be called **dual metric**. A complete dual metric locally convex space is called an $(F')$-**space**.

**0.7.6.** The relation between metrizable and dual metric locally convex spaces is made precise by the two following propositions.

**Proposition.** *The topological dual $E'_b$ of each metrizable locally convex space $E$ is an $(F')$-space.*

**Proposition.** *The topological dual $E'_b$ of each dual metric locally convex space $E$ is an $(F)$-space.*

## 0.8. Banach Spaces

**0.8.1.** A semi-norm $p_U$ on a real or complex linear space $E$ is said to be a **norm** if $p_U(x) > 0$ for $x \neq o$. Each norm $p_U$ yields a topology on $E$ if the sets $\varepsilon U$ with $\varepsilon > 0$ and

$$U = \{x \in E : p_U(x) \leqq 1\}$$

are used as a fundamental system of zero neighborhoods. A locally convex space $E$ so constructed is called **normed** or **normable** and $U$ is called the **closed unit ball** of $E$. A complete normed space is called a **Banach space**.

The strong topological dual $E'_b$ of each normed space $E$ is a Banach space with the norm

$$p'_U(a) = \sup \{|\langle x, a\rangle| : x \in U\}$$

and closed unit ball $U^0$. We then have

$$p_{U^0}(a) = p'_U(a) \quad \text{for } a \in E' .$$

**0.8.2.** We have the following answer to the question of what hypotheses imply the topology on a locally convex space is derived from a norm.

**Kolmogorov Theorem.** *A locally convex space $E$ is normable if and only if it has a bounded zero neighborhood.*

**0.8.3.** As a complement to the normability criterion above we formulate the following

**Proposition.** *A locally convex space $E$ is finite dimensional if and only if it has a precompact zero neighborhood.*

## 0.9. Hilbert Spaces

**0.9.1.** A mapping which determines for each pair of elements $x$ and $y$ from a linear space $E$ over the field $K$ of real or complex numbers a number $(x, y) \in K$ is called a **semi-scalar product** if it has the following properties:

($S_1$) *For $x, y, z \in E$ and $\alpha, \beta \in K$ we have the relation*

$$(\alpha x + \beta y, z) = \alpha(x, z) + \beta(y, z) .$$

($S_2$) *For $x, y \in E$ we have $(x, y) = \overline{(y, x)}$.*

($S_3$) *For $x \in E$ it always follows that $(x, x) \geqq 0$.*

If in addition $(x, x) = 0$ always implies $x = o$, then the given mapping is called a **scalar product**.

For each semi-scalar product one obtains a semi-norm $p_U$ by setting

$$p_U(x) = (x, x)^{1/2} \quad \text{for} \quad x \in E .$$

**0.9.2.** A **Hilbert space** is a real or complex Banach space whose norm is obtained from a scalar product.

**0.9.3. Riesz Representation Theorem.** *When $E$ is a Hilbert space, each linear form $a \in E'$ can be represented by a uniquely determined element $y \in E$ by means of the equation*

$$\langle x, a\rangle = (x, y) \quad \textit{for } x \in E .$$

*In this case we have*

$$p_{U^\circ}(a) = (y, y)^{1/2} .$$

**0.9.4.** Two elements $x$ and $y$ of a Hilbert space $E$ are called **orthogonal** if $(x, y) = 0$. An orthonormal system $[e_i, I]$ is a set of elements $e_i$ which correspond in a definite way to the elements of an index set $I$ and for which the relations

$$(e_i, e_j) = \delta_{ij}$$

hold. For each orthonormal system $[e_i, I]$ we have the **Bessel inequality**

$$\sum_I |(x, e_i)|^2 \leqq (x, x) \quad \text{for } x \in E .$$

The definition of the sum on the left is found in 1.1. For an arbitrary linear form $a \in E'$ the Bessel inequality takes the following form because of the Riesz representation theorem

$$\sum_I |\langle e_i, a\rangle|^2 \leqq p_{U^\circ}(a)^2 .$$

An orthonormal system $[e_i, I]$ is called **complete** if each element $x \in E$ can be represented in the sense of 1.3 as the sum

$$x = \sum_I (x, y_i)\, e_i .$$

**Proposition.** *In each Hilbert space there exists a complete orthonormal system.*

## 0.10. Continuous Linear Mappings in Locally Convex Spaces

**0.10.1.** If $E$ and $F$ are two arbitrary real or complex locally convex spaces a mapping $T$ from $E$ into $F$ is called **linear** if for $x, y \in E$ and $\alpha, \beta \in K$ we always have the relation

$$T(\alpha x + \beta y) = \alpha T x + \beta T y .$$

The collection $\mathfrak{R}(T)$ of all elements $T x$ with $x \in E$ is called the **range** of $T$.

A linear mapping $T$ is **continuous** if and only if for each zero neighborhood $V \in \mathfrak{U}(F)$ there is a zero neighborhood $U \in \mathfrak{U}(E)$ with $T(U) \subset V$.

**0.10.2.** Each linear mapping $T$ from $E$ into $F$ determines a linear mapping $T'$ from $F'$ into $E'$ whose value on the linear form $b \in F'$ is the linear form $a \in E'$ defined by the equation

$$\langle x, a \rangle = \langle T x, b \rangle \quad \text{for } x \in E .$$

$T'$ is called the **dual mapping** corresponding to $T$. It can be shown that the mapping $T'$ is continuous in the strong or weak topology on $E'$ and $F'$.

When $E$ and $F$ are two Hilbert spaces, then we can consider instead of the dual mapping $T'$, the **adjoint mapping** $T^*$ which is defined in the following way. For each element $y \in E$, $T^* y$ is the element $z \in E$ with

$$(x, z) = (T x, y) \quad \text{for } x \in E ,$$

which is uniquely determined by the Riesz Representation Theorem.

**0.10.3.** For two locally convex spaces $E$ and $F$, the collection of all continuous linear mappings from $E$ into $F$ is a linear space $\mathcal{L}(E, F)$ if for $S, T \in \mathcal{L}(E, F)$ and $\alpha, \beta \in K$, the continuous linear mapping $\alpha S + \beta T$ is defined by the equation

$$(\alpha S + \beta T) x = \alpha S x + \beta T x \quad \text{for } x \in E .$$

**0.10.4.** A topology on the linear space $\mathcal{L}(E, F)$ can be constructed from the semi-norms

$$p_{(A,V)}(T) = \sup \{ |\langle T x, b \rangle| : x \in A, b \in V^0 \} \text{ with } A \in \mathfrak{B}(E) \text{ and } V \in \mathfrak{U}(F).$$

The locally convet space obtained in this way is denoted by $\mathcal{L}_b(E, F)$.

When $E$ and $F$ are two normed spaces with closed unit balls $U$ and $V$ then the topology of $\mathcal{L}_b(E, F)$ is obtained from the norm

$$\beta(T) = \inf \{ \varrho > 0 : T(U) \subset \varrho V \} .$$

**Proposition.** *If $E$ is a normed space and $F$ a Banach space, then $\mathcal{L}_b(E, F)$ is a Banach space.*

**0.10.5.** For two locally convex spaces $E$ and $F$ a mapping $T \in \mathcal{L}(E, F)$ is called **finite** if its image space is finite dimensional. Each mapping of this kind can be represented in the form

$$T x = \sum_{r=1}^{n} \langle x, a_r \rangle y_r \quad \text{for } x \in E$$

with linear forms $a_1, \ldots, a_n \in E'$ and elements $y_1, \ldots, y_n \in F$.

The collection of all finite mappings is a linear subspace of $\mathcal{L}(E, F)$ which is denoted by $\mathcal{A}(E, F)$.

**0.10.6.** A linear mapping $T$ from a normed space $E$ into a normed space $F$ is called **precompact** if the image $T(U)$ of the closed unit ball $U$ of $E$ is a precompact subset of $F$.

**Proposition.** *Each precompact linear mapping has a separable range.*

**Proposition.** *Each mapping $T \in \mathcal{L}(E, F)$ for which there is a directed system of finite mappings $T_\alpha \in \mathcal{A}(E, F)$ with*

$$\lim_{\alpha} \boldsymbol{\beta}(T - T_\alpha) = 0$$

*is precompact.*

A linear mapping $T$ from $E$ into $F$ is called **compact** if there is a compact subset $K$ of $F$ with $T(U) \subset K$.

**Proposition.** *Each precompact linear mapping $T$ from a normed space $E$ into a Banach space $F$ is compact.*

Moreover, we have the

**Proposition.** *The dual mapping $T'$ of each precompact linear mapping $T$ is compact.*

**0.10.7.** For $E$ a Hilbert space, a mapping $P \in \mathcal{L}(E, E)$ is called a **projection** when the statements

$$P = P^* \quad \text{and} \quad P^2 = P$$

are valid. The norm $\boldsymbol{\beta}(P)$ of a projection $P \neq 0$ is 1.

**Proposition.** *For each closed linear subspace $H$ of a Hilbert space $E$ there is a projection $P$ with $\mathfrak{R}(P) = H$.*

## 0.11. The Normed Spaces Associated with a Locally Convex Space

**0.11.1.** If $U$ is an arbitrary closed and absolutely convex zero neighborhood of the locally convex space $E$, then we set

$$N(U) = \{x \in E : p_U(x) = 0\}$$

and denote by $x(U)$ the equivalence class corresponding to the element $x \in E$ in the quotient space $E(U) = E/N(U)$. We shall require that the topology determined by the norm

$$p[x(U)] = p_U(x) \qquad \text{for } x \in E$$

always be used on the linear space $E(U)$.

**0.11.2.** For each closed and absolutely convex bounded subset $A$ of a locally convex space $E$

$$E(A) = \{x \in E : x \in \varrho A \text{ for some } \varrho > 0\}$$

is a linear subspace of $E$ on which we can construct a topology from the norm

$$p_A(x) = \inf \{\varrho > 0 : x \in \varrho A\} \qquad \text{for } x \in E(A) .$$

**0.11.3.** For $U \in \mathfrak{U}(E)$ we can determine a continuous linear form $\mathfrak{a}$ on $E(U)$ for each linear form $a \in E'(U^0)$ by means of the equation

$$\langle x(U), \mathfrak{a}\rangle = \langle x, a\rangle \qquad \text{for } x \in E,$$

where the norm of $\mathfrak{a}$ is given by

$$p'(\mathfrak{a}) = \sup \{|\langle x, a\rangle| : x \in U\} = p_{U^0}(a) .$$

Since all linear forms $\mathfrak{a} \in [E(U)]'_b$ can be obtained in this way, we can identify the topological dual $[E(U)]'_b$ with the Banach space $E'(U^0)$ constructed in 0.11.2.

**0.11.4.** If $A$ is an arbitrary set in $\mathfrak{B}(E)$, then for each linear form $a \in E'$ a linear form $\mathfrak{a} \in [E(A)]'$ is defined by the equation

$$\langle x, \mathfrak{a}\rangle = \langle x, a\rangle \qquad \text{for } x \in E(A) ,$$

where the norm of $\mathfrak{a}$ is given by

$$p'(\mathfrak{a}) = \sup \{|\langle x, a\rangle| : x \in A\} = p_{A^0}(a) = p[a(A^0)] .$$

The normed space $E'(A^0)$ formed in the sense of 0.11.1 can thus be considered as a linear subspace of the topological dual $[E(A)]'_b$ by identifying the equivalence class $a(A^0)$ with the linear form $\mathfrak{a}$.

Moreover, the normed space $E(A)$ can also be considered as a linear subspace of the topological dual $[E'(A^0)]'_b$. For this we must associate with each element $x \in E(A)$ the linear form $\mathfrak{x} \in [E'(A^0)]'$, which is defined by the equation

$$\langle \mathfrak{x}, a(A^0)\rangle = \langle x, a\rangle \qquad \text{for } a \in E'$$

and has the norm

$$p'(\mathfrak{x}) = \sup \{|\langle x, a\rangle| : a \in A^0\} = p_A(x) .$$

**0.11.5.** If $M$ and $N$ are two absolutely convex subsets of $E$ for which the relation $N \subset \varrho M$ holds for some positive number $\varrho$, then we write $N \prec M$ (read: $M$ absorbs $N$, $N$ is absorbed by $M$). It is clear that the relation $\prec$ thus defined is transitive.

For two zero neighbourhoods $U, V \in \mathfrak{U}(E)$ with $V \prec U$, we can define a **canonical mapping** $E(V, U)$ from $E(V)$ to $E(U)$ by associating with each equivalence class $x(V) \in E(V)$ the uniquely determined equivalence class $x(U) \in E(U)$.

On the other hand, if we have two bounded subsets $A, B \in \mathfrak{B}(E)$ with $A \prec B$ then $E(A) \subset E(B)$ and the **canonical mapping** $E(A, B)$ from $E(A)$ into $E(B)$ is defined by the equation

$$E(A, B)\, x = x \qquad \text{for } x \in E(A)\,.$$

Finally, for $A \in \mathfrak{B}(E)$ and $U \in \mathfrak{U}(E)$ it is always true that $A \prec U$ and we can define a **canonical mapping** $E(A, U)$ from $E(A)$ into $E(U)$ by setting

$$E(A, U)\, x = x(U) \qquad \text{for } x \in E(A)\,.$$

## 0.12. Radon Measures

**0.12.1.** The collection of all continuous real or complex functions $\varphi = [\varphi(x)]$ which are defined on a compact Hausdorff space $M$ is a linear space $\mathscr{C}(M)$ with respect to the operations

$$[\varphi(x)] + [\psi(x)] = [\varphi(x) + \psi(x)] \qquad \text{and} \qquad \alpha[\varphi(x)] = [\alpha\, \varphi(x)]$$

which becomes a Banach space by the introduction of the norm

$$\|\varphi\| = \sup\,\{|\varphi(x)| : x \in M\}\,.$$

Moreover, $\mathscr{C}(M)$ has the property that for each $\varphi = [\varphi(x)] \in \mathscr{C}(M)$, it always follows that $|\varphi| = [|\varphi(x)|] \in \mathscr{C}(M)$.

**0.12.2.** The linear forms $\mu$ on $\mathscr{C}(M)$ are called **Radon measures**. A Radon measure $\mu$ is called **positive** if for each function $\varphi \in \mathscr{C}(M)$ with $\varphi(x) \geqq 0$ we have $\langle \varphi, \mu \rangle \geqq 0$. For an arbitrary positive Radon measure $\mu$ we have the inequality

$$|\langle \varphi, \mu \rangle| \leqq \langle |\varphi|, \mu \rangle \qquad \text{for all} \quad \varphi \in \mathscr{C}(M)\,,$$

and for the norm of $\mu$, which is denoted by $\mu(M)$, we have the relation

$$\mu(M) = \langle [1], \mu \rangle\,.$$

**0.12.3.** For each Radon measure $\mu$ we obtain by means of the equation

$$\langle \varphi, |\mu| \rangle = \sup\,\{|\langle \psi, \mu \rangle| : \psi \in \mathscr{C}(M) \quad \text{with } |\psi(x)| \leqq \varphi(x)\}$$

a real valued function $|\mu|$ on the set $\mathscr{C}^+(M)$ of all functions $\varphi \in \mathscr{C}(M)$ with $\varphi(x) \geqq 0$ which has the following properties:

$$\langle \varphi + \psi, |\mu| \rangle = \langle \varphi, |\mu| \rangle + \langle \psi, |\mu| \rangle \quad \text{for} \quad \varphi, \psi \in \mathscr{C}^+(M)\,,$$

$$\langle \alpha\, \varphi, |\mu| \rangle = \alpha \langle \varphi, |\mu| \rangle \quad \text{for} \quad \varphi \in \mathscr{C}^+(M) \text{ and } \alpha \geqq 0\,.$$

This function can be extended uniquely to a positive Radon measure which we denote by $|\mu|$ also. It is always true that

$$|\langle \varphi, \mu \rangle| \leqq \langle |\varphi|, |\mu| \rangle \quad \text{for } \varphi \in \mathscr{C}(M)\,.$$

**0.12.4.** For each positive Radon measure $\mu$, there is a uniquely determined $\sigma$-ring $\mathfrak{R}_\mu(M)$ of so-called $\mu$-**measurable subsets** of the compact Hausdorff space $M$. To each set $A \in \mathfrak{R}_\mu(M)$ there corresponds a non-negative number $\mu(A)$ called the **measure** of $A$. Since the value of $\mu$ for all continuous functions $\varphi$ can be calculated by an integration process from the associated set function, we shall in the sequel usually write the number $\langle \varphi, \mu \rangle$ in the form

$$\int_M \varphi(x)\, d\mu\,.$$

**0.12.5.** For each positive Radon measure $\mu$, a semi-scalar product is obtained on $\mathscr{C}(M)$ by the equation

$$(\varphi, \psi) = \int_M \varphi(x)\, \overline{\psi(x)}\, d\mu$$

and the associated semi-norm is

$$\boldsymbol{\lambda}_2(\varphi) = \left\{\int_M |\varphi(x)|^2\, d\mu\right\}^{1/2}.$$

The Hilbert space $\mathfrak{L}^2_\mu(M)$ whose elements can be interpreted as equivalence classes $\hat{f}$ of real or complex functions $f$ arises by completing the quotient space of $\mathscr{C}(M)$ by the linear subspace $\{\varphi \in \mathscr{C}(M) : \boldsymbol{\lambda}_2(\varphi) = 0\}$ formed in the sense of 0.11.1. Two functions belong to the same equivalence class if and only if they differ from one another only on a set $A \in \mathfrak{R}_\mu(M)$ with $\mu(A) = 0$.

**0.12.6.** The function $f_A$ with

$$f_A(x) = 1 \quad \text{for } x \in A \quad \text{and} \quad f_A(x) = 0 \quad \text{for } x \notin A$$

is called the **characteristic function** of the subset $A$. It can be shown that the equivalence classes $\hat{f}$ which correspond to the so-called $\mu$-**step functions**

$$f(x) = \sum_{r=1}^{n} \alpha_r\, f_{A_r}(x) \quad \text{with} \quad A_r \in \mathfrak{R}_\mu(M)$$

form a dense linear subspace of $\mathfrak{L}^2_\mu(M)$.

Chapter I

# Summable Families

In 1931 W. Orlicz [1] studied unconditionally convergent infinite series in Banach spaces and also introduced the concept of weak unconditional convergence. However, not until 1950 was it proven that in each infinite dimensional normed space there is an unconditionally convergent series which does not converge absolutely (Theorem of Dvoretzky and Rogers 3.4.1).

The definition of *summable families of numbers* given in 1.1.1 is due to E. H. Moore who was able to show that an infinite series of real or complex numbers converges unconditionally if and only if it is summable. The great advantage of this definition is that it can also be applied to uncountable families. The study of families in normed spaces was undertaken by T. H. Hildebrandt.

In this chapter we deal with *weakly summable, summable* and *absolutely summable families* in an arbitrary locally convex space. Here it will be shown that these families form linear spaces on which useful locally convex topologies can be constructed.

The *property* $(B)$ introduced in 1.5.5 has great significance. In Theorem 4.3.1, for instance, we shall prove that only those nuclear locally convex spaces which have this property also possess a nuclear strong topological dual.

## 1.1. Summable Families of Numbers

**1.1.1.** A **family of numbers** $[\xi_i, I]$ is a set of real or complex numbers $\xi_i$ which are corresponded in an unique way to the elements $i$ of an index set $I$.

If we denote the collection of all finite subsets $\mathfrak{i}$ of $I$ by $\mathfrak{F}(I)$, then for each family of numbers $[\xi_i, I]$, the finite **partial sums**

$$\sigma_{\mathfrak{i}} = \sum_{\mathfrak{i}} \xi_i \quad \text{with } \mathfrak{i} \in \mathfrak{F}(I)$$

form a directed system with set theoretic inclusion $\mathfrak{j} \supset \mathfrak{i}$ used as the $\geqq$-relation.

If the directed system $\{\sigma_{\mathfrak{i}}\}$ converges, then the family of numbers $[\xi_i, I]$ is called **summable**. The uniquely determined limit $\sigma$ is then designated as the **sum** of the family of numbers and we write

$$\sigma = \sum_I \xi_i .$$

If the index set $I$ is finite, then $\sigma$ obviously coincides with the ordinary sum.

It must be emphasized that the order of the numbers $\xi_i$ plays no part in determining $\sigma$.

**1.1.2. Lemma.** *If for the family of numbers $[\xi_i, I]$ there is a positive number $\varrho$ with*

$$|\sum_{\mathfrak{i}} \xi_i| \leqq \varrho \qquad \text{for all } \mathfrak{i} \in \mathfrak{F}(I) ,$$

*we then have the inequality*

$$\sum_{\mathfrak{i}} |\xi_i| \leqq 4\varrho \qquad \text{for all } \mathfrak{i} \in \mathfrak{F}(I) .$$

*Proof.* We first assume that the family of numbers $[\xi_i, I]$ is real. Then each set $\mathfrak{i} \in \mathfrak{F}(I)$ can be divided into two disjoint parts:

$$\mathfrak{i}_+ = \{i \in \mathfrak{i} : \xi_i \geqq 0\} \qquad \text{and} \qquad \mathfrak{i}_- = \{i \in \mathfrak{i} : \xi_i < 0\} .$$

Since

$$\sum_{\mathfrak{i}_+} |\xi_i| = \sum_{\mathfrak{i}_+} \xi_i \leqq \varrho \qquad \text{and} \qquad \sum_{\mathfrak{i}_-} |\xi_i| = - \sum_{\mathfrak{i}_-} \xi_i \leqq \varrho ,$$

we then have the inequality

$$\sum_{\mathfrak{i}} |\xi_i| = \sum_{\mathfrak{i}_+} |\xi_i| + \sum_{\mathfrak{i}_-} |\xi_i| \leqq 2\varrho .$$

If the family of numbers $[\xi_i, I]$ is complex, then for the real families of numbers $[\mathrm{Re}(\xi_i), I]$ and $[\mathrm{Im}(\xi_i), I]$, we have the inequalities

$$\sum_{\mathfrak{i}} |\mathrm{Re}(\xi_i)| \leqq 2\varrho \qquad \text{and} \qquad \sum_{\mathfrak{i}} |\mathrm{Im}(\xi_i)| \leqq 2\varrho$$

because

$$\left|\sum_{\mathfrak{i}} \mathrm{Re}(\xi_i)\right| = \left|\mathrm{Re}\left(\sum_{\mathfrak{i}} \xi_i\right)\right| \leqq \varrho \quad \text{and} \quad \left|\sum_{\mathfrak{i}} \mathrm{Im}(\xi_i)\right| = \left|\mathrm{Im}\left(\sum_{\mathfrak{i}} \xi_i\right)\right| \leqq \varrho .$$

Consequently we have

$$\sum_{\mathfrak{i}} |\xi_i| \leqq 4\varrho .$$

**1.1.3. Theorem.** *A family of numbers $[\xi_i, I]$ is summable if and only if there is a positive number $\varrho$ for which the inequality*

$$\sum_{\mathfrak{i}} |\xi_i| \leqq \varrho \qquad \text{for all } \mathfrak{i} \in \mathfrak{F}(I)$$

*is valid.*

*Proof.* If $[\xi_i, I]$ is a family of numbers with the sum $\sigma$, we determine a set $\mathfrak{i}_0 \in \mathfrak{F}(I)$ such that

$$|\sigma_{\mathfrak{j}} - \sigma| \leqq 1 \quad \text{for all } \mathfrak{j} \in \mathfrak{F}(I) \quad \text{with } \mathfrak{j} \supset \mathfrak{i}_0 .$$

Then we have the equation

$$\left|\sum_{\mathfrak{i}} \xi_i\right| = \left|\sum_{\mathfrak{i} \cup \mathfrak{i}_0} \xi_i - \sigma + \sigma - \sum_{\mathfrak{i}_0 \setminus \mathfrak{i}} \xi_i\right| \leqq 1 + |\sigma| + \left|\sum_{\mathfrak{i}_0} \xi_i\right| .$$

for each set $\mathfrak{i} \in \mathfrak{F}(I)$. Therefore, by Lemma 1.1.2 we have for each $\mathfrak{i} \in \mathfrak{F}(I)$ the inequality

$$\sum_{\mathfrak{i}} |\xi_i| \leqq 4\,(1 + |\sigma| + \sum_{\mathfrak{i}_0} |\xi_i|) ,$$

in which the right hand side is constant.

We now consider a family of numbers $[\xi_i, I]$, for which there is a positive number $\varrho$ with

$$\sum_{\mathfrak{i}} |\xi_i| \leqq \varrho \qquad \text{for all } \mathfrak{i} \in \mathfrak{F}(I) .$$

Then we have the inequality

$$\varrho_0 = \sup\left\{\sum_{\mathfrak{i}} |\xi_i| : \mathfrak{i} \in \mathfrak{F}(I)\right\} \leqq \varrho < +\infty . \tag{1}$$

Thus there exists a monotonically increasing sequence $\mathfrak{i}_1, \mathfrak{i}_2, \ldots$ of sets from $\mathfrak{F}(I)$ with

$$\sum_{\mathfrak{i}_n} |\xi_i| \geqq \varrho_0 - \frac{1}{n} . \tag{2}$$

Because of (1) and (2) we have the inequality

$$\sum_{\mathfrak{i}} |\xi_i| \leqq \frac{1}{n} \qquad \text{for all } \mathfrak{i} \in \mathfrak{F}(I) \text{ with } \mathfrak{i} \cap \mathfrak{i}_n = \emptyset . \tag{3}$$

If set we

$$\sigma_n = \sum_{\mathfrak{i}_n} \xi_i ,$$

then because of (3) we have the relation

$$|\sigma_m - \sigma_n| \leqq \sum_{\mathfrak{i}_m \setminus \mathfrak{i}_n} |\xi_i| \leqq \frac{1}{n}$$

for all natural numbers $m$ and $n$ with $m \geqq n$. Consequently, $\{\sigma_n\}$ is a Cauchy sequence which must have a limit $\sigma$. We can thus determine for each positive number $\delta$ a natural number $n$ with $n \geqq 2/\delta$ and $|\sigma_n - \sigma| \leqq \delta/2$. Then, for all sets $\mathfrak{i} \in \mathfrak{F}(I)$ with $\mathfrak{i} \supset \mathfrak{i}_n$ we have the inequality

$$|\sigma_{\mathfrak{i}} - \sigma| = \left|\sum_{\mathfrak{i} \setminus \mathfrak{i}_n} \xi_i + \sigma_n - \sigma\right| \leqq 1/n + \delta/2 \leqq \delta .$$

Therefore, we have shown that the directed system of finite partial sums converges to $\sigma$.

**1.1.4.** From Theorem 1.1.3 it follows that when the family of numbers $[\xi_i, I]$ is summable, so is $[|\xi_i|, I]$. Thus the limit

$$\sum_I |\xi_i| = \sup\left\{ \sum_{\mathfrak{i}} |\xi_i| : \mathfrak{i} \in \mathfrak{F}(I) \right\}$$

exists. Hence we can express the fact that the family of numbers $[\xi_i, I]$ is summable by the inequality

$$\sum_I |\xi_i| < +\infty .$$

**1.1.5. Proposition.** *Each summable family of numbers contains at most countably many non-vanishing terms.*

*Proof.* With the notation of 1.1.3 we set

$$I_0 = \bigcup_{n=1}^{\infty} \mathfrak{i}_n .$$

Then $I_0$ is finite or countably infinite and for each index $i \in I \backslash I_0$ we have the inequality $|\xi_i| \leqq 1/n$ for each natural number $n$ because $\{i\} \cap \mathfrak{i}_n = \emptyset$ on account of (3). Therefore, we have

$$\xi_i = 0 \qquad \text{for } i \in I \backslash I_0 .$$

**1.1.6.** A family of numbers $[\alpha_i, I]$ is said to **converge to 0** if for each positive number $\delta$ there is a set $\mathfrak{i} \in \mathfrak{F}(I)$ such that the inequality

$$|\alpha_i| \leqq \delta \qquad \text{for all } i \notin \mathfrak{i}$$

is valid.

The collection $\boldsymbol{c}_I$ of these families of numbers forms a Banach space with respect to the operations

$$[\alpha_i, I] + [\beta_i, I] = [\alpha_i + \beta_i, I] \qquad \text{and} \qquad \lambda\,[\alpha_i, I] = [\lambda\,\alpha_i, I]$$

with the norm

$$\lambda_\infty[\alpha_i, I] = \sup\,\{|\alpha_i| : i \in I\} .$$

Since each continuous linear form $x$ on $\boldsymbol{c}_I$ can be represented in the form

$$\langle [\alpha_i, I], x\rangle = \sum_I \alpha_i\, \xi_i$$

where $[\xi_i, I]$ is a uniquely determined summable family of numbers, it follows that the topological dual of $\boldsymbol{c}_I$ coincides with the Banach space $\boldsymbol{l}_I^1$ of all summable families of numbers. Here the norm

$$\lambda_1[\xi_i, I] = \sum_I |\xi_i|$$

is used on $\boldsymbol{l}_I^1$.

It follows in the same way that the topological dual of $\boldsymbol{l}_I^1$ can be identified with the Banach space $\boldsymbol{m}_I$ of all bounded families of numbers.

To supplement these remarks we formulate the

**Lemma.** *If* $[\xi_i, I]$ *is a family of numbers such that*

$$\sum_I |\alpha_i \xi_i| < +\infty$$

*for each* $[\alpha_i, I] \in \boldsymbol{c}_I$ *we then have the inequality*

$$\sum_I |\xi_i| < +\infty .$$

*Proof.* If the relation

$$\sum_I |\xi_i| = +\infty$$

held for the family $[\xi_i, I]$ we could then determine a monotonically increasing sequence of sets $\mathfrak{i}_r \in \mathfrak{F}(I)$ beginning with $\mathfrak{i}_0 = \emptyset$ such that the inequalities

$$\sum_{\mathfrak{i}_r} |\xi_i| \geqq r^2$$

would be valid. Then the family of numbers $[\alpha_i, I]$ with $\alpha_i = 1/n$ for $i \in \mathfrak{i}_n \backslash \mathfrak{i}_{n-1}$ and $n = 1, 2, \ldots$ and $\alpha_i = 0$ for $i \notin \bigcup_{n=1}^{\infty} \mathfrak{i}_n$ would belong to $\boldsymbol{c}_I$ and the inequality

$$\sum_I |\alpha_i \xi_i| < +\infty$$

would have to hold. But this is impossible since

$$\sum_I |\alpha_i \xi_i| \geqq \sum_{\mathfrak{i}_r} |\alpha_i \xi_i| \geqq r \qquad \text{for } r = 0, 1, \ldots .$$

**1.1.7.** A family of numbers $[\xi_i, I]$ is called **square summable** if the inequality

$$\sum_I |\xi_i|^2 < +\infty$$

is valid.

The collection $\boldsymbol{l}_I^2$ of these families of numbers forms a linear space with respect to the operations

$$[\xi_i, I] + [\eta_i, I] = [\xi_i + \eta_i, I] \qquad \text{and} \qquad \lambda[\xi_i, I] = [\lambda\, \xi_i, I] ,$$

on which a scalar product can be introduced by means of

$$([\xi_i, I], [\eta_i, I]) = \sum_I \xi_i \bar{\eta}_i .$$

Here the sum on the right hand side exists for each pair of square summable families of numbers because of the **Hölder Inequality**

$$\sum_I |\xi_i \eta_i| \leqq \Big\{\sum_I |\xi_i|^2\Big\}^{1/2} \Big\{\sum_I |\eta_i|^2\Big\}^{1/2} < +\infty .$$

It can be shown that $\mathfrak{l}_I^2$ is complete in the topology obtained from the associated norm

$$\lambda_2[\xi_i, I] = \left\{\sum_I |\xi_i|^2\right\}^{1/2}.$$

Therefore $\mathfrak{l}_I^2$ is a Hilbert space.

We mention the following

**Lemma.** *If* $[\alpha_i, I]$ *is a family of numbers such that*

$$\sum_I |\alpha_i \xi_i| \leqq \alpha \left\{\sum_I |\xi_i|^2\right\}^{1/2}$$

*for each* $[\xi_i, I] \in \mathfrak{l}_I^2$ *we then have the inequality*

$$\sum_I |\alpha_i|^2 \leqq \alpha^2 .$$

## 1.2. Weakly Summable Families in Locally Convex Spaces

**1.2.1.** By a family $[x_i, I]$ from a locally convex space $E$ we understand a set of elements $x_i \in E$ which correspond in a unique way to the elements $i$ of an index set $I$.

Such a family $[x_i, I]$ is called **weakly summable** if for each continuous linear form $a \in E'$ we have the inequality

$$\sum_I |\langle x_i, a\rangle| < +\infty .$$

**1.2.2.** From this definition it immediately follows that the collection $\mathfrak{l}_I^1[E]$ of all weakly summable families $[x_i, I]$ from $E$ forms a linear space with respect to the operations

$$[x_i, I] + [y_i, I] = [x_i + y_i, I] \quad \text{and} \quad \alpha[x_i, I] = [\alpha x_i, I] .$$

**1.2.3.** For each family $[x_i, I] \in \mathfrak{l}_I^1[E]$

$$A = \left\{\sum_{\mathfrak{i}} \alpha_i x_i : |\alpha_i| \leqq 1 \quad \text{and} \quad \mathfrak{i} \in \mathfrak{F}(I)\right\}$$

is a weakly bounded subset of $E$ because for each continuous linear form $a \in E'$ we have the relation

$$\left|\left\langle \sum_{\mathfrak{i}} \alpha_i x_i, a\right\rangle\right| \leqq \sum_I |\langle x_i, a\rangle| < +\infty .$$

Consequently, by 0.6.3 $A$ must be a bounded subset of $E$. For each zero neighborhood $U \in \mathfrak{U}(E)$ there is thus a positive number $\varrho$ with

$$\sum_{\mathfrak{i}} \alpha_i x_i \in \varrho U \quad \text{for } |\alpha_i| \leqq 1 \quad \text{and} \quad \mathfrak{i} \in \mathfrak{F}(I) .$$

In particular if we determine for each continuous linear form $a \in U^0$ the numbers $\alpha_i$ with $|\alpha_i| = 1$ such that $\alpha_i \langle x_i, a\rangle = |\langle x_i, a\rangle|$, we then obtain the equation

$$\sum_{\mathfrak{i}} |\langle x_i, a\rangle| = \left\langle \sum_{\mathfrak{i}} \alpha_i\, x_i, a\right\rangle \leqq \varrho \quad \text{for } \mathfrak{i} \in \mathfrak{F}(I) \text{ and } a \in U^0 .$$

We therefore have

$$\boldsymbol{\varepsilon}_U[x_i, I] = \sup\left\{\sum_I |\langle x_i, a\rangle| : a \in U^0\right\} \leqq \varrho < +\infty .$$

It is easy to see that a system of semi-norms $\boldsymbol{\varepsilon}_U[x_i, I]$ on $\boldsymbol{l}_I^1[E]$ can be obtained by means of these equations. The topology which is obtained from these semi-norms in the usual way will be designated as the **$\boldsymbol{\varepsilon}$-topology**.

It is clear that for constructing this $\boldsymbol{\varepsilon}$-topology we do not need all semi-norms $\boldsymbol{\varepsilon}_U[x_i, I]$ with $U \in \mathfrak{U}(E)$. Rather, it suffices to let $U$ range over an arbitrary fundamental system of zero neighborhoods.

**1.2.4. Proposition.** *For each complete locally convex space E the space $\boldsymbol{l}_I^1[E]$ is also complete.*

*Proof.* We consider in $\boldsymbol{l}_I^1[E]$ an arbitrary directed Cauchy-system $[x_i^{[\alpha]}, I]$. Then, for each index $i \in I$, $\{x_i^{[\alpha]}\}$ is a directed Cauchy-system in $E$ and there is an element $x_i \in E$ with

$$\lim_\alpha x_i^{[\alpha]} = x_i . \tag{1}$$

We now show that the given directed Cauchy-system converges in the $\boldsymbol{\varepsilon}$-topology to the family $[x_i, I]$ formed from the elements $x_i$. For this purpose we determine for each zero neighborhood $U \in \mathfrak{U}(E)$ a $\beta_0$ with

$$\sum_I |\langle x_i^{[\alpha]} - x_i^{[\beta]}, a\rangle| \leqq 1 \quad \text{for } a \in U^0 \text{ and } \alpha, \beta \geqq \beta_0 . \tag{2}$$

Because of (1) we have by passing to the limit in (2)

$$\sum_I |\langle x_i - x_i^{[\beta]}, a\rangle| \leqq 1 \quad \text{for } a \in U^0 \text{ and } \beta \geqq \beta_0 . \tag{3}$$

Hence for each $a \in U^0$ we have the inequality

$$\begin{aligned}\sum_I |\langle x_i, a\rangle| &\leqq \sum_I |\langle x_i - x_i^{[\beta]}, a\rangle| + \sum_I |\langle x_i^{[\beta]}, a\rangle| \\ &\leqq 1 + \sum_I |\langle x_i^{[\beta]}, a\rangle| < +\infty .\end{aligned}$$

Since each continuous linear form $a \in E'$ belongs to $U^0$ for some zero neighborhood $U \in \mathfrak{U}(E)$ it follows that the family $[x_i, I]$ belongs to $\boldsymbol{l}_I^1[E]$. Because of (3) we have the relation

$$\boldsymbol{\varepsilon} - \lim_\beta\, [x_i^{[\beta]}, I] = [x_i, I] .$$

The completeness of $\boldsymbol{l}_I^1[E]$ is thus established.

## 1.3. Summable Families in Locally Convex Spaces

**1.3.1.** For each weakly summable family $[x_i, I] \in \mathfrak{l}_I^1[E]$ and all sets $\mathfrak{i} \in \mathfrak{F}(I)$, the families $[x_i(\mathfrak{i}), I]$ with $x_i(\mathfrak{i}) = x_i$ for $i \in \mathfrak{i}$ and $x_i(\mathfrak{i}) = o$ for $i \notin \mathfrak{i}$ also belong to $\mathfrak{l}_I^1[E]$. These families form a directed system when set theoretic inclusion $\mathfrak{j} \supset \mathfrak{i}$ is used as the $\geqq$-relation.

In the sequel we shall designate a family $[x_i, I] \in \mathfrak{l}_I^1[E]$ as **summable** if the following relation holds

$$\varepsilon - \lim_{\mathfrak{i}} [x_i(\mathfrak{i}), I] = [x_i, I] .$$

**1.3.2.** It follows immediately from this definition that the collection $\mathfrak{l}_I^1(E)$ of all summable families $[x_i, I]$ from $E$ form a linear space with respect to the operations

$$[x_i, I] + [y_i, I] = [x_i + y_i, I] \quad \text{and} \quad \alpha [x_i, I] = [\alpha x_i, I] .$$

On this linear space we shall always use the $\varepsilon$-topology induced by $\mathfrak{l}_I^1[E]$.

**1.3.3. Proposition.** *For each locally convex space, $\mathfrak{l}_I^1(E)$ is a closed linear subspace of $\mathfrak{l}_I^1[E]$.*

*Proof*. We consider a family $[x_i, I] \in \mathfrak{l}_I^1[E]$, which lies in the closed hull of $\mathfrak{l}_I^1(E)$. Then for each zero neighborhood $U \in \mathfrak{U}(E)$ there is a family $[y_i, I] \in \mathfrak{l}_I^1(E)$ with

$$\varepsilon_U [x_i - y_i, I] \leqq 1/3 .$$

We now determine a set $\mathfrak{i}_0 \in \mathfrak{F}(I)$ such that

$$\varepsilon_U [y_i - y_i(\mathfrak{i}), I] \leqq 1/3 \quad \text{for all } \mathfrak{i} \in \mathfrak{F}(I) \quad \text{with } \mathfrak{i} \supset \mathfrak{i}_0 .$$

Then since

$$\varepsilon_U[x_i - x_i(\mathfrak{i}), I] \leqq \varepsilon_U[x_i - y_i, I] + \varepsilon_U[y_i - y_i(\mathfrak{i}), I] + \varepsilon_U [y_i(\mathfrak{i}) - x_i(\mathfrak{i}), I]$$

we have the inequality

$$\varepsilon_U[x_i - x_i(\mathfrak{i}), I] \leqq 1 \quad \text{for all } \mathfrak{i} \in \mathfrak{F}(I) \text{ with } \mathfrak{i} \subset \mathfrak{i}_0 .$$

Consequently we obtain

$$\varepsilon - \lim_{\mathfrak{i}} [x_i(\mathfrak{i}), I] = [x_i, I] ,$$

and we have shown that the family $[x_i, I]$ lies in $\mathfrak{l}_I^1(E)$. Thus our assertion is proved.

**1.3.4.** Because of 1.2.4 and 1.3.3 we have

**Proposition.** *For each complete locally convex space, $\mathfrak{l}_I^1(E)$ is also complete.*

**1.3.5.** We note without proof that in a weakly quasi-complete locally convex space each weak summable family is summable. In general, however, $\mathfrak{l}^1_I(E)$ is a proper linear subspace of $\mathfrak{l}^1_I[E]$. Moreover, we have the

**Proposition.** *From* $[x_i, I] \in \mathfrak{l}^1_I[E]$ *and* $[\alpha_i, I] \in \mathfrak{c}_I$ *it always follows that* $[\alpha_i\, x_i, I] \in \mathfrak{l}^1_I(E)$.

*Proof.* If $U$ is an arbitrary zero neighborhood from $\mathfrak{U}(E)$ we then determine a set $\mathfrak{i}_0 \in \mathfrak{F}(I)$ with

$$|\alpha_i| \leqq (\varepsilon_U[x_i, I] + 1)^{-1} \quad \text{for} \quad i \notin \mathfrak{i}_0 .$$

Then for each set $\mathfrak{i} \in \mathfrak{F}(I)$ with $\mathfrak{i} \supset \mathfrak{i}_0$ and all $a \in U^0$ we have the inequality

$$\sum_{I \setminus \mathfrak{i}} |\langle \alpha_i\, x_i, a\rangle| \leqq (\varepsilon_U[x_i, I] + 1)^{-1} \sum_I |\langle x_i, a\rangle| \leqq 1$$

which establishes the relation

$$\varepsilon - \lim_i\, [\alpha_i\, x_i(\mathfrak{i}), I] = [\alpha_i\, x_i, I] .$$

**1.3.6. Theorem.** *A family* $[x_i, I]$ *from a locally convex space* $E$ *is summable if and only if the finite partial sums*

$$s_{\mathfrak{i}} = \sum_{\mathfrak{i}} x_i \quad \textit{with} \quad \mathfrak{i} \in \mathfrak{F}(I)$$

*form a directed Cauchy-system.*

*Proof.* If $\{s_{\mathfrak{i}}\}$ is a directed Cauchy-system, then for each zero neighborhood $U \in \mathfrak{U}(E)$ we can determine a set $\mathfrak{i}_0 \in \mathfrak{F}(I)$ such that

$$s_{\mathfrak{i}_1} - s_{\mathfrak{i}_2} \in (1/4)\, U \quad \text{for all} \quad \mathfrak{i}_1, \mathfrak{i}_2 \in \mathfrak{F}(I) \quad \text{with} \quad \mathfrak{i}_1, \mathfrak{i}_2 \supset \mathfrak{i}_0 .$$

Thus for $a \in U^0$ and $\mathfrak{j} \in \mathfrak{F}(I \setminus \mathfrak{i}_0)$, we have the inequality

$$\left|\sum_{\mathfrak{j}} \langle x_i, a\rangle\right| = |\langle s_{\mathfrak{j} \cup \mathfrak{i}_0} - s_{\mathfrak{i}_0}, a\rangle| \leqq 1/4 ,$$

from which we obtain the relation

$$\sum_{I \setminus \mathfrak{i}_0} |\langle x_i, a\rangle| = \sup\left\{\sum_{\mathfrak{j}} |\langle x_i, a\rangle| : \mathfrak{j} \in \mathfrak{F}(I \setminus \mathfrak{i}_0)\right\} \leqq 1 \quad \text{for} \quad a \in U^0$$

by reason of Lemma 1.1.2. But this means that for each set $\mathfrak{i} \in \mathfrak{F}(I)$ with $\mathfrak{i} \supset \mathfrak{i}_0$ the assertion $\varepsilon_U[x_i - x_i(\mathfrak{i}), I] \leqq 1$ is valid. Since $U$ is an arbitrary zero neighborhood from $\mathfrak{U}(E)$, the relation

$$\varepsilon - \lim_{\mathfrak{i}}\, [x_i(\mathfrak{i}), I] = [x_i, I] \quad \text{or} \quad [x_i, I] \in \mathfrak{l}^1_I(E)$$

is thereby proved.

On the other hand, if $[x_i, I]$ is an arbitrary summable family from $E$, the assertion

$$\varepsilon - \lim_{\mathfrak{i}}\, [x_i(\mathfrak{i}), I] = [x_i, I] ,$$

is then valid, and the families $[x_i(\mathfrak{i}), I]$ form a directed Cauchy-system in $l_I^1(E)$. Hence, for each zero neighborhood $U \in \mathfrak{U}(E)$, there is a set $\mathfrak{i}_0 \in \mathfrak{F}(I)$ such that

$$\varepsilon_U[x_i(\mathfrak{i}_1) - x_i(\mathfrak{i}_2), I] \leqq 1 \quad \text{for } \mathfrak{i}_1, \mathfrak{i}_2 \in \mathfrak{F}(I) \text{ with } \mathfrak{i}_1, \mathfrak{i}_2 \supset \mathfrak{i}_0$$

is valid. Because of this, we have for all continuous linear forms $a \in U^0$ the inequality

$$|\langle s_{\mathfrak{i}_1} - s_{\mathfrak{i}_2}, a\rangle| = \left|\left\langle \sum_I [x_i(\mathfrak{i}_1) - x_i(\mathfrak{i}_2)], a\right\rangle\right| \leqq \sum_I |\langle x_i(\mathfrak{i}_1) - x_i(\mathfrak{i}_2), a\rangle|$$
$$\leq \varepsilon_U[x_i(\mathfrak{i}_1) - x_i(\mathfrak{i}_2), I] \leqq 1 ,$$

so that the relation

$$s_{\mathfrak{i}_1} - s_{\mathfrak{i}_2} \in U \quad \text{for } \mathfrak{i}_1, \mathfrak{i}_2 \in \mathfrak{F}(I) \text{ with } \mathfrak{i}_1, \mathfrak{i}_2 \supset \mathfrak{i}_0$$

holds because of the Bipolar theorem. We have thus shown that the finite partial sums $s_\mathfrak{i}$ form a directed Cauchy-system for each summable family $[x_i, I]$.

**1.3.7.** Because of the theorem just proved we are able to define the **sum** of each family $[x_i, I] \in l_I^1(E)$ to be the limit $s$ of the directed Cauchy-system $\{s_i\}$ which lies in the complete hull of $E$. We then write

$$s = \sum_I x_i .$$

If the index set $I$ is finite, then $s$ evidently coincides with the usual sum.

It must be emphasized that the ordering of the elements $x_i$ plays no role in determining $s$.

## 1.4. Absolutely Summable Families in Locally Convex Spaces

**1.4.1.** A family $[x_i, I]$ from a locally convex space $E$ is called **absolutely summable** if for each zero neighborhood $U \in \mathfrak{U}(E)$, we have the inequality

$$\sum_I p_U(x_i) < +\infty .$$

**1.4.2.** It follows immediately from this definition that the collection $l_I^1\{E\}$ of all absolutely summable families $[x_i, I]$ from $E$ forms a linear space with respect to the operations

$$[x_i, I] + [y_i, I] = [x_i + y_i, I] \quad \text{and} \quad \alpha[x_i, I] = [\alpha x_i, I] .$$

A locally convex topology, which we shall designate as the **$\pi$-topology**, can be constructed on the linear space $l_I^1\{E\}$ from the semi-norms

$$\pi_U[x_i, I] = \sum_I p_U(x_i) \quad \text{with } U \in \mathfrak{U}(E) .$$

It is clear that for constructing this $\boldsymbol{\pi}$-topology it is not necessary to use all semi-norms $\boldsymbol{\pi}_U[x_i, I]$ with $U \in \mathfrak{U}(E)$. Rather, it suffices for $U$ to range over an arbitrary fundamental system of zero neighborhoods.

**1.4.3. Proposition.** *For each complete locally convex space, $\boldsymbol{l}_I^1\{E\}$ is also complete.*

*Proof.* We consider an arbitrary directed Cauchy-system $\{[x_i^{[\alpha]}, I]\}$ in $\boldsymbol{l}_I^1\{E\}$. Then for each index $i \in I$, $\{x_i^{[\alpha]}\}$ is a directed Cauchy-system in $E$ and there is an element $x_i \in E$ with

$$\lim_{\alpha} x_i^{[\alpha]} = x_i \,. \tag{1}$$

We now show that the given directed Cauchy-system converges in the $\boldsymbol{\pi}$-topology to the family $[x_i, I]$ formed from the elements $x_i$. For this purpose we determine for each zero neighborhood $U \in \mathfrak{U}(E)$ a $\beta_0$ with

$$\sum_I p_U(x_i^{[\alpha]} - x_i^{[\beta]}) \leqq 1 \quad \text{for} \quad \alpha, \beta \geqq \beta_0 \,. \tag{2}$$

Because of (1) we have by passing to the limit in (2)

$$\sum_I p_U(x_i - x_i^{[\beta]}) \leqq 1 \quad \text{for} \quad \beta \geqq \beta_0 \,. \tag{3}$$

Consequently, for each zero neighborhood $U \in \mathfrak{U}(E)$ we have the inequality

$$\sum_I p_U(x_i) \leqq \sum_I p_U(x_i - x_i^{[\beta]}) + \sum_I p_U(x_i^{[\beta]}) \leqq 1 + \sum_I p_U(x_i^{[\beta]}) < +\infty \,,$$

so that the family $[x_i, I]$ belongs to $\boldsymbol{l}_I^1\{E\}$. The relation

$$\boldsymbol{\pi} - \lim_{\alpha} [x_i^{[\alpha]}, I] = [x_i, I]$$

now follows immediately from (3). The completeness of $\boldsymbol{l}_I^1\{E\}$ is thus proved.

**1.4.4.** As the analog to the definition of summable family presented in 1.3.1 we obtain the

**Proposition.** *For each absolutely summable family $[x_i, I]$ from a locally convex space $E$,*

$$\boldsymbol{\pi} - \lim_{\mathfrak{i}} [x_i(\mathfrak{i}), I] = [x_i, I] \,.$$

*Proof.* If $U$ is an arbitrary zero neighborhood from $\mathfrak{U}(E)$ then, since

$$\sum_I p_U(x_i) < +\infty \,,$$

there is a set $\mathfrak{i}_0 \in \mathfrak{F}(I)$ with

$$\sum_{I \setminus \mathfrak{i}_0} p_U(x_i) \leqq 1 \,.$$

Consequently, we have

$$\boldsymbol{\pi}_U[x_i - x_i(\mathfrak{i}), I] \leqq 1 \quad \text{for all } \mathfrak{i} \in \mathfrak{F}(I) \text{ with } \mathfrak{i} \supset \mathfrak{i}_0,$$

and the desired relation is proved.

**1.4.5. Proposition.** *For each locally convex space E, we have the relation*

$$\boldsymbol{l}_I^1\{E\} \subset \boldsymbol{l}_I^1(E),$$

*and the identity mapping from $\boldsymbol{l}_I^1\{E\}$ into $\boldsymbol{l}_I^1(E)$ is continuous.*

*Proof.* We consider an arbitrary absolutely summable family $[x_i, I]$ from the locally convex space $E$. Since for each continuous linear form $a \in E'$ there is a zero neighborhood $U \in \mathfrak{U}(E)$ with $a \in U^0$, we always have

$$\sum_I |\langle x_i, a\rangle| \leqq \sum_I p_U(x_i) < +\infty,$$

and the family $[x_i, I]$ is weakly summable. This leads to the relation

$$\boldsymbol{l}_I^1\{E\} \subset \boldsymbol{l}_I^1[E].$$

Since for each zero neighborhood $U \in \mathfrak{U}(E)$ and all families $[x_i, I] \in \boldsymbol{l}_I^1\{E\}$ the relation

$$\sum_I |\langle x_i, a\rangle| \leqq \sum_I p_U(x_i) \quad \text{with} \quad a \in U^0$$

is valid, we have

$$\boldsymbol{\varepsilon}_U[x_i, I] \leqq \boldsymbol{\pi}_U[x_i, I] \quad \text{for} \quad [x_i, I] \in \boldsymbol{l}_I^1\{E\}.$$

Therefore, the identity mapping from $\boldsymbol{l}_I^1\{E\}$ into $\boldsymbol{l}_I^1[E]$ is continuous. It is now seen that each absolutely summable family is also summable since the relation

$$\boldsymbol{\pi} - \lim_{\mathfrak{i}} [x_i(\mathfrak{i}), I] = [x_i, I]$$

proved in 1.4.4 implies

$$\boldsymbol{\varepsilon} - \lim_{\mathfrak{i}} [x_i(\mathfrak{i}), I] = [x_i, I].$$

## 1.5. Totally Summable Families in Locally Convex Spaces

**1.5.1.** A family $[x_i, I]$ from a locally convex space $E$ is called **totally summable** if there is a bounded subset $B \in \mathfrak{B}(E)$ for which the inequality

$$\sum_I p_B(x_i) < +\infty$$

holds.

**1.5.2.** It is an immediately consequence of this definition that the collection $\boldsymbol{l}_I^1\langle E\rangle$ of all totally summable families $[x_i, I]$ from $E$ forms a linear space with respect to the operations

$$[x_i, I] + [y_i, I] = [x_i + y_i, I] \quad \text{and} \quad \alpha[x_i, I] = [\alpha\, x_i, I]\,.$$

**1.5.3. Proposition.** *Each totally summable family $[x_i, I]$ from a locally convex space $E$ contains at most countably many elements $x_i \neq o$.*

*Proof.* The inequality

$$\sum_I p_B(x_i) < +\infty$$

and Proposition 1.1.5 imply that the set

$$I_0 = \{i \in I : p_B(x_i) > 0\} = \{i \in I : x_i \neq o\}$$

is at most countably infinite.

**1.5.4. Proposition.** *Each totally summable family from a locally convex space is absolutely summable.*

*Proof.* For each totally summable family $[x_i, I]$ from a locally convex space $E$ there is by definition a bounded subset $B \in \mathfrak{B}(E)$ with

$$\sum_I p_B(x_i) < +\infty\,.$$

However, for each zero neighborhood $U \in \mathfrak{U}(E)$, there is a positive number $\varrho$ with $B \subset \varrho\, U$ so that the inequality

$$\sum_I p_U(x_i) \leqq \varrho \sum_I p_B(x_i) < +\infty\,.$$

holds.

**1.5.5.** We say that a locally convex space $E$ has **property** $(B)$ if for each bounded subset $\mathbf{B}$ of $\boldsymbol{l}_N^1\{E\}$ there is a set $B \in \mathfrak{B}(E)$ such that

$$\sum_N p_B(x_n) \leqq 1 \quad \text{for} \quad [x_n, N] \in \mathbf{B}\,.$$

Here, $N$ stands for the set of natural numbers.

**1.5.6. Proposition.** *In each locally convex space $E$ with property $(B)$, all absolutely summable families are also totally summable.*

*Proof.* We consider in $E$ an arbitrary absolutely summable family $[x_i, I]$. If $\mathfrak{i} = \{i_1, \ldots, i_k\}$ is a finite subset of $I$ we set $x_n(\mathfrak{i}) = x_{i_n}$ for $n = 1, \ldots, k$ and $x_n(\mathfrak{i}) = o$ for $n > k$. Then the families $[x_n(\mathfrak{i}), N]$ with $\mathfrak{i} \in \mathfrak{F}(I)$ form a bounded subset of $\boldsymbol{l}_N^1\{E\}$ because

$$\boldsymbol{\pi}_U[x_n(\mathfrak{i}), N] = \sum_{\mathfrak{i}} p_U(x) \leqq \boldsymbol{\pi}_U[x_i, I] \quad \text{for} \quad U \in \mathfrak{U}(E)\,,$$

and by hypothesis we can find a set $B \in \mathfrak{B}(E)$ with

$$\sum_{\mathfrak{i}} p_B(x_i) = \sum_N p_B(x_n(\mathfrak{i})) \leqq 1 \quad \text{for} \quad \mathfrak{i} \in \mathfrak{F}(I)\,.$$

It follows that

$$\sum_I p_B(x_i) \leqq 1 .$$

**1.5.7.** We now present an *example of a locally convex space which does not have property* (*B*).

If $I$ is an arbitrary index set we consider the linear space $\Omega_I$ of all families of numbers $[\xi_i, I]$ and endow it with the locally convex topology determined by the semi-norms

$$p_{\mathfrak{i}}[\xi_i, I] = \sum_{\mathfrak{i}} |\xi_i|,$$

where $\mathfrak{i}$ ranges over all finite subsets of $I$. Then the family $[e_j, J]$ with $J = I$ formed from the families of numbers $e_j = [\delta_i^{[j]}, I]$ is absolutely summable. However, since $e_j \neq 0$ for each $j$, $\Omega_I$ can never have property ($B$) for an uncountable index set $I$ because of 1.5.3 and 1.5.6.

**1.5.8. Theorem.** *Each metric or dual metric locally convex space $E$ has property* ($B$).

*Proof.*

(1) For the case in which the locally convex space $E$ is metric, we consider a countable fundamental system of zero neighborhoods $U_1, U_2, \ldots$. Then, for each bounded subset $\mathbf{B}$ of $\mathfrak{l}_I^1\{E\}$ there is a positive number $\varrho_n$ with

$$\sum_I p_{U_n}(x_i) \leqq \varrho_n \quad \text{for } [x_i, I] \in \mathbf{B} .$$

If we form the set

$$B = \left\{x \in E : \sum_N \frac{1}{2^n \varrho_n} p_{U_n}(x) \leqq 1\right\},$$

which belongs to $\mathfrak{B}(E)$ we then have the inequality

$$\sum_I p_B(x_i) = \sum_N \frac{1}{2^n \varrho_n} \sum_I p_{U_n}(x_i) \leqq 1 \quad \text{for all } [x_i, I] \in \mathbf{B}$$

because

$$p_B(x) = \sum_N \frac{1}{2^n \varrho_n} p_{U_n}(x) \quad \text{for } x \in E(B) .$$

(2) If the locally convex space $E$ is dual metric then we consider a countable fundamental system of bounded subsets $B_1 \subset B_2 \subset \ldots$. Now let $\mathbf{B}$ be a bounded subset of $\mathfrak{l}_I^1\{E\}$ for which we shall assume that the assertion

$$\sigma_n = \sup\left\{\sum_I p_{B_n}(x_i) : [x_i, I] \in \mathbf{B}\right\} = +\infty$$

holds for all natural numbers $n$. Then there are families $[x_i^{[n]}, I] \in \mathbf{B}$ and sets $\mathfrak{i}_n \in \mathfrak{F}(I)$ with

$$\sum_{\mathfrak{i}_n} p_{B_n}(x_i^{[n]}) > 2^n .$$

Thus there exist for $i \in \mathfrak{i}_n$ and $n = 1, 2, \ldots$ continuous linear forms $a_i^{[n]} \in B_n^0$ with

$$\sum_{\mathfrak{i}_n} |\langle x_i^{[n]}, a_i^{[n]} \rangle| > 2^n .$$

We now determine for the bounded subset $B_m$ a positive number $\varrho_m \geqq 1$ with

$$a_i^{[n]} \in \varrho_m B_m^0 \quad \text{for} \quad i \in \mathfrak{i}_n \quad \text{and} \quad n = 1, \ldots, m .$$

Since the relation

$$a_i^{[n]} \in B_n^0 \subset B_m^0 \subset \varrho_m B_m^0$$

holds for $i \in \mathfrak{i}_n$ and $n > m$, we have

$$A = \{a_i^{[n]} : i \in \mathfrak{i}_n \quad \text{and} \quad n = 1, 2, \ldots\} \subset \varrho_m B_m^0 .$$

Consequently, $A$ is a countable and strongly bounded subset of $E'$ which by hypothesis must be equicontinuous. There is thus a zero neighborhood $U \in \mathfrak{U}(E)$ with $A \subset U^0$.

Since **B** is supposed to be a bounded of $\boldsymbol{l}_I^1\{E\}$, there exists a positive number $\varrho$ with

$$\sum_I p_U(x_i) \leqq \varrho \qquad \text{for} \quad [x_i, I] \in \mathbf{B} .$$

This is impossible because it would lead to the inequality

$$k \leqq \sum_{n=1}^{k} \frac{1}{2^n} \sum_{\mathfrak{i}_n} |\langle x_i^{[n]}, a_i^{[n]} \rangle| \leqq \sum_N \frac{1}{2^n} \sum_I p_U(x_i^{[n]}) \leqq \varrho$$

for all natural numbers $k$. From this contradiction it follows that for at least one natural number $n$ the inequality

$$\sigma_n = \sup\left\{\sum_I p_{B_n}(x_i) : [x_i, I] \in \mathbf{B}\right\} < +\infty$$

must hold. If we now set $B = \sigma_n B_n$ we obtain

$$\sum_I p_B(x_i) \leqq 1 \qquad \text{for all} \quad [x_i, I] \in \mathbf{B} .$$

**1.5.9.** It is easy to see that property (B) remains valid when we pass to linear subspaces of a locally convex space. However, the space $\Omega_I$ for uncountable index set does not have property $(B)$ even though it can be represented as the quotient space of a locally convex space with property $(B)$ (Y. Kōmura).

## 1.6. Finite Dimensional Families in Locally Convex Spaces

**1.6.1.** A family $[x_i, I]$ from a locally convex space $E$ is called **finite dimensional** if all elements $x_i$ lie in a finite dimensional subspace.

**1.6.2. Proposition.** *Each finite dimensional weakly summable family from a locally convex space is also totally summable.*

*Proof.* If $[x_i, I]$ is such a family in a locally convex space $E$, then there are finitely many linearly independent elements $e_1, \ldots, e_n$ of which the elements $x_i$ are linear combinations:

$$x_i = \sum_{r=1}^{n} \xi_i^{[r]} e_r \quad \text{for } i \in I .$$

By the Hahn Banach Theorem we can find continuous linear forms $f_1, \ldots, f_n \in E'$ with $\langle e_r, f_s \rangle = \delta_{rs}$. We then have the relations

$$\xi_i^{[r]} = \langle x_i, f_r \rangle \quad \text{for } i \in I \text{ and } r = 1, \ldots, n ,$$

from which the inequalities

$$\sum_I |\xi_i^{[r]}| = \sum_I |\langle x_i, f_r \rangle| < +\infty$$

follow for $r = 1, \ldots, n$.

Finally, if $B$ is a bounded subset from $\mathfrak{B}(E)$ with $e_1, \ldots, e_n \in B$, we have

$$\sum_I p_B(x_i) \leqq \sum_{r=1}^{n} \left( \sum_I |\xi_i^{[r]}| \right) p_B(e_r) < +\infty .$$

Chapter 2

# Absolutely Summing Mappings

Let $E$ and $F$ be two normed spaces; we call a continuous linear mapping $T$ from $E$ into $F$ *absolutely summing* if it takes each summable family $[x_n, N]$ from $E$ into an absolutely summable family $[T x_n, N]$. Such mappings were first considered by A. Pełczyński [2] and A. Pietsch [4]. However, they also implicitly in the book of A. Grothendieck ([3], Chap. I, p. 155) as "applications semi-integrales à droite".

In the sequel it is shown that the collection of all absolutely summing mappings $T \in \mathcal{L}(E, F)$ forms a linear space $\mathcal{P}(E, F)$ on which a useful norm can be introduced. The principal item of this chapter is Theorem 2.2.3 of A. Pietsch [4] in which a characterization of absolutely summing mappings is presented and which has far reaching consequences. The canonical mapping from $\boldsymbol{l}_N^1$ into $\boldsymbol{l}_N^2$ is introduced as the simplest non-trivial example of an absolutely summing mapping (A. Pełczyński et W. Szlenk [1] and A. Pietsch [4]).

Especially noteworthy is the fact, established implicitly by A. Grothendieck [5], [6], that the absolutely summing mappings in Hilbert space coincide with those named for D. Hilbert and E. Schmidt, which can be represented by an infinite matrix with

$$\sum_{I, J} |\alpha_{ij}|^2 < +\infty$$

(Theorem 2.5.5).

## 2.1. Absolutely Summing Mappings in Locally Convex Spaces

**2.1.1.** Let $E$ and $F$ be two arbitrary locally convex spaces. A continuous linear mapping $T$ from $E$ into $F$ is called **absolutely summing** if it takes each summable family $[x_n, N]$ from $E$ into an absolutely summable family $[T x_n, N]$ from $F$. Here $N$ is the set of natural numbers.

**2.1.2. Theorem.** *If $I$ is an arbitrary index set then for each absolutely summing mapping $T \in \mathcal{L}(E, F)$ a linear mapping from $\boldsymbol{l}_I^1[E]$ into $\boldsymbol{l}_I^1\{F\}$ is defined by the equation*

$$T_I[x_i, I] = [T x_i, I],$$

*and this mapping $T_I$ maps bounded subsets of $\mathfrak{l}_I^1[E]$ into bounded subsets of $\mathfrak{l}_I^1\{F\}$.*

*Proof.* We consider a bounded subset $\mathbf{B}$ of $\mathfrak{l}_I^1[E]$ and assume that for some zero neighborhood $V \in \mathfrak{U}(F)$ the assertion

$$\sigma_V = \sup\left\{\sum_I p_V(T\,x_i) : [x_i, I] \in \mathbf{B}\right\} = +\infty$$

is valid. Then there are families $[x_i^{[n]}, I] \in \mathbf{B}$ and sets $\mathfrak{i}_n \in \mathfrak{F}(I)$ with

$$\sum_{\mathfrak{i}_n} p_V(T\,x_i^{[n]}) > 2^n \quad \text{for} \quad n = 1, 2, \ldots.$$

Since $\mathbf{B}$ is bounded there exists for each linear form $a \in E'$ a positive number $\varrho$ with

$$\sum_I |\langle x_i, a\rangle| \leqq \varrho \quad \text{for all} \quad [x_i, I] \in \mathbf{B}\,,$$

so that we have the inequality

$$\sum_N \sum_{\mathfrak{i}_n} \left|\left\langle \frac{1}{2^n}\,x_i^{[n]}, a\right\rangle\right| \leqq \varrho < +\infty \quad \text{for all} \quad a \in E'\,.$$

Consequently the family $\left[\frac{1}{2^n}\,x_i^{[n]}, P\right]$ with the countable index set

$$P = \{(i, n) : i \in \mathfrak{i}_n \quad \text{and} \quad n = 1, 2, \ldots\}$$

is weakly summable, and the families

$$\left[\frac{1}{2^n}\alpha_i^{[n]}\,x_i^n, P\right] \quad \text{with} \quad [\alpha_i^{[n]}, P] \in \mathfrak{c}_P\,,$$

which by 1.3.5 are summable, are carried by $T$ into absolutely summable families. Consequently we have

$$\sum_N \sum_{\mathfrak{i}_n} \frac{1}{2^n}\,|\alpha_i^{[n]}|\,p_V(T\,x_i^{[n]}) < +\infty \quad \text{for all} \quad [\alpha_i^{[n]}, P] \in \mathfrak{c}_P\,.$$

On the basis of Lemma 1.1.6 this leads to the false assertion

$$\sum_N \sum_{\mathfrak{i}_n} \frac{1}{2^n}\,p_V(T\,x_i^{[n]}) < +\infty\,.$$

From this contradiction follows that for each zero neighborhood $V \in \mathfrak{U}(F)$ there is a positive number $\sigma_V$ with

$$\sum_I p_V(T\,x_i) \leqq \sigma_V \quad \text{for all} \quad [x_i, I] \in \mathbf{B}\,.$$

Thus $\mathbf{B}$ is mapped into a bounded subset of $\mathfrak{l}_I^1\{F\}$ by the mapping $T_I$.

In particular, since each weakly summable family $[x_i, I]$ can be considered as a bounded singleton of $\mathfrak{l}_I^1[E]$, every family $[T\,x_i, I]$ with $[x_i, I] \in \mathfrak{l}_I^1[E]$ is absolutely summable.

**2.1.3.** As an refinement of the statement of 2.1.2 we have the

**Theorem.** *For each absolutely summing mapping $T$ from a metric locally convex space $E$ into an arbitrary locally convex space $F$, a continuous linear mapping from $\boldsymbol{l}_I^1[E]$ into $\boldsymbol{l}_I^1\{F\}$ is defined by the equation*

$$T_I[x_i, I] = [T\, x_i, I]\,.$$

*Proof.* For each zero neighborhood $V \in \mathfrak{U}(F)$

$$\mathbf{U} = \left\{[x_i, I] \in \boldsymbol{l}_I^1[E] : \sum_I p_V(T\, x_i) \leqq 1\right\}$$

is a closed and absolutely convex subset of $\boldsymbol{l}_I^1[E]$ which, because of the previous theorem, absorbs every bounded subset. Since $\boldsymbol{l}_I^1[E]$ as a metric locally convex space is quasi-barrelled, $\mathbf{U}$ must be a zero neighborhood in $\boldsymbol{l}_I^1[E]$. The required continuity of $T_I$ is thus established.

**2.1.4.** If the linear space $\boldsymbol{l}_K^1$ is provided with the locally convex topology obtained from the semi-norms

$$\boldsymbol{\lambda}_A[\xi_k, K] = \sum_A |\xi_k| + \sup_K |\xi_k|\,.$$

($A$ ranges over all countable subsets of $K$), we then obtain a dual metric locally convex space $E$, and the identity mapping $T$ of $E$ into the Banach space $F = \boldsymbol{l}_K^\infty$ is absolutely summing. When the index set $K$ is taken to be uncountable the mapping defined by the equation

$$T_I[x_i, I] = [T\, x_i, I]$$

is not continuous from $\boldsymbol{l}_I^1[E]$ into $\boldsymbol{l}_I^1\{F\}$. (Y. Kōmura).

## 2.2. Absolutely Summing Mappings in Normed Spaces

**2.2.1.** In this section we shall always assume that $E$ and $F$ are two normed spaces with closed unit balls $U$ and $V$.

**Proposition.** *A continuous linear mapping $T$ from $E$ into $F$ is absolutely summing if and only if there is a number $\varrho \geqq 0$ such that the relation*

$$\sum_{\mathfrak{n}} p(T\, x_n) \leqq \varrho \sup\left\{\sum_{\mathfrak{n}} |\langle x_n, a\rangle| : a \in U^0\right\}$$

*holds for all finite families $[x_n, \mathfrak{n}]$ from $E$.*

*Proof.* By Theorem 2.1.2 there is for each absolutely summing mapping $T$ a positive number $\varrho$, such that for all families $[x_n, N] \in \boldsymbol{l}_N^1[E]$ with $\boldsymbol{\varepsilon}_U[x_n, N] \leqq 1$ the relation $\boldsymbol{\pi}_V[T\, x_n, N] \leqq \varrho$ is valid. Consequently we have

$$\boldsymbol{\pi}_V[T\, x_n, N] \leqq \varrho\, \boldsymbol{\varepsilon}_U[x_n, N] \quad \text{for} \quad [x_n, N] \in \boldsymbol{l}_N^1[E]\,.$$

Since each finite family $[x_n, \mathfrak{n}]$ can be extended to a family $[x_n, N]$ by setting $x_n = 0$ for $n \notin \mathfrak{n}$, the necessity of the given condition is thus established.

Conversely, if for the continuous linear mapping $T$ there is a number $\varrho$ with the required property, then for each finite subfamily $[x_n, \mathfrak{n}]$ of a summable family $[x_n, N]$ we have the relation

$$\sum_{\mathfrak{n}} p_V(T x_n) \leqq \varrho \sup \left\{\sum_{\mathfrak{n}} |\langle x_n, a\rangle| : a \in U^0\right\} \leqq \varrho\, \varepsilon_U[x_n, N] .$$

Thus we get

$$\sum_{\mathfrak{n}} p_V(T x_n) = \sup \left\{\sum_{\mathfrak{n}} p_V(T x_n) : \mathfrak{n} \in \mathfrak{F}(N)\right\} < +\infty ,$$

and we have shown that $T$ carries each summable family into an absolutely summable family.

**2.2.2.** If $T$ is an arbitrary absolutely summing mapping from $E$ into $F$ then $\boldsymbol{\pi}(T)$ will mean the infimum of all numbers $\varrho$ which have the property given in the preceeding theorem. It is easy to see that $\boldsymbol{\pi}(T)$ coincides with the norm $\boldsymbol{\beta}(T_N)$ of the mapping $T_N$ and that for all weakly summable families $[x_i, I]$ over an arbitrary index set we have the relation

$$\boldsymbol{\pi}_V[T x_i, I] \leqq \boldsymbol{\pi}(T)\, \boldsymbol{\varepsilon}_U[x_i, I] .$$

**2.2.3.** If the collection of all absolutely summing mappings from $E$ into $F$ is denoted by $\mathscr{P}(E, F)$ we then obtain the

**Proposition.** *$\mathscr{P}(E, F)$ is a linear space with the norm $\boldsymbol{\pi}(T)$.*

*Proof.* Our conclusion follows immediately from the following relations:

$$\boldsymbol{\beta}(T) \leqq \boldsymbol{\pi}(T) , \tag{0}$$

$$\boldsymbol{\pi}(S + T) = \boldsymbol{\beta}(S_N + T_N) \leqq \boldsymbol{\beta}(S_N) + \boldsymbol{\beta}(T_N) = \boldsymbol{\pi}(S) + \boldsymbol{\pi}(T) \tag{1}$$

$$\boldsymbol{\pi}(\alpha\, T) = \boldsymbol{\beta}(\alpha\, T_N) = |\alpha|\, \boldsymbol{\beta}(T_N) = |\alpha|\, \boldsymbol{\pi}(T) . \tag{2}$$

Here $S$ and $T$ are supposed to be two arbitrary absolutely summing mappings from $E$ into $F$; and $\alpha$, a real or complex number.

**2.2.4. Lemma.** *If $\{T_\alpha\}$ is a directed Cauchy-system from $\mathscr{P}(E, F)$ and there is a mapping $T \in \mathscr{L}(E, F)$ with $\lim_\alpha T_\alpha x = T x$ for $x \in E$, then $T$ belongs to $\mathscr{P}(E, F)$ and we have $\boldsymbol{\pi} - \lim_\alpha T_\alpha = T$.*

*Proof.* For an arbitrary given positive number $\delta$ we determine $\beta_0$ with

$$\boldsymbol{\pi}(T_\alpha - T_\beta) \leqq \delta \quad \text{for } \alpha, \beta \geqq \beta_0 .$$

Then for each finite family $[x_n, \mathfrak{n}]$ from $E$ we have the inequality

$$\sum_{\mathfrak{n}} p_V(T_\alpha x_n - T_\beta x_n) \leqq \delta \sup \left\{\sum_{\mathfrak{n}} |\langle x_n, a\rangle| : a \in U^0\right\},$$

from which by passing to the limit we obtain the relation

$$\sum_{\mathfrak{n}} p_V(T\,x_n - T_\beta\,x_n) \leqq \delta \sup\left\{\sum_{\mathfrak{n}} |\langle x_n, a\rangle| : a \in U^0\right\}$$

for $\beta \geqq \beta_0$. Consequently the mapping $T - T_{\beta_0}$, and thus also $T$, belongs to $\mathscr{P}(E, F)$. Moreover, since the relation $\boldsymbol{\pi}(T - T_\beta) \leqq \delta$ holds for all $\beta \geqq \beta_0$, the statement $\boldsymbol{\pi} - \lim_\alpha T_\alpha = T$ is also proved.

As a simple consequence of the Lemma just proved we have the

**Proposition.** *For a normed space E and a Banach space F, $\mathscr{P}(E, F)$ is a Banach space.*

**2.2.5.** If $E$, $F$ and $G$ are three normed spaces with closed unit balls $U$, $V$ and $W$ we then obtain the

**Proposition.**

(1) *If* $T \in \mathscr{L}(E, F)$ *and* $S \in \mathscr{P}(F, G)$, *then* $S\,T \in \mathscr{P}(E, G)$ *and* $\boldsymbol{\pi}(S\,T) \leqq \boldsymbol{\pi}(S)\,\boldsymbol{\beta}(T)$.

(2) *If* $T \in \mathscr{P}(E, F)$ *and* $S \in \mathscr{L}(F, G)$, *then* $S\,T \in \mathscr{P}(E, G)$ *and* $\pi)S\,T) \leqq \boldsymbol{\beta}(S)\,\boldsymbol{\pi}(T)$.

*Proof.* Our assertion results from the following inequalities which are valid for each finite family $[x_n, \mathfrak{n}]$ because of Proposition 2.2.1:

$$\sum_{\mathfrak{n}} p_W(S\,T\,x_n) \leqq \boldsymbol{\pi}(S) \sup\left\{\sum_{\mathfrak{n}} |\langle x_n, T'\,b\rangle| : b \in V^0\right\}$$

$$\leqq \boldsymbol{\pi}(S)\,\boldsymbol{\beta}(T) \sup\left\{\sum_{\mathfrak{n}} |\langle x_n, a\rangle| : a \in U^0\right\}, \qquad (1)$$

$$\sum_{\mathfrak{n}} p_W(S\,T\,x_n) \leqq \boldsymbol{\beta}(S) \sum_{\mathfrak{n}} p_V(T\,x_n)$$

$$\leqq \boldsymbol{\beta}(S)\,\boldsymbol{\pi}(T) \sup\left\{\sum_{\mathfrak{n}} |\langle x_n, a\rangle| : a \in U^0\right\}. \qquad (2)$$

## 2.3. A Characterization of Absolutely Summing Mappings in Normed Spaces

**2.3.1.** Let $E$ be a normed space with closed unit ball $U$; a weakly compact subset $M$ of $U^0$ is called **essential** if the relation

$$p_U(x) = \varrho \sup\{|\langle x, a\rangle| : a \in M\} \qquad \text{for all} \quad x \in E$$

is valid.

It is obvious that the entire set $U^0$ is always essential. There are, however, cases when $M$ can be chosen much smaller. We consider for example the Banach space $\mathscr{C}(M)$ of continuous functions on a compact

Hausdorff space $M$ for which the set of all Dirac measures $\delta_x$ with $x \in M$ defined by the equation

$$\langle \varphi, \delta_x \rangle = \varphi(x) \quad \text{for} \quad \varphi \in \mathscr{C}(M)$$

is essential.

**2.3.2.** For the real unit interval or the complex unit circle $\Delta$ the $I$-fold topological product $\Delta_I$ is compact by the Tichonov Theorem 0.1.10. Moreover, if $M$ is an arbitrary essential subset of $E'$, then $\Delta_I \times M$ as the topological product of $\Delta_I$ and $M$ must be a compact Hausdorff space.

From these preliminary remarks we derive the following

**Lemma.** *For each family* $[x_i, I] \in \mathfrak{l}_I^1(E)$ *the function* $\Phi$ *with*

$$\Phi(\alpha_i, a) = \sum_I \alpha_i \langle x_i, a \rangle$$

*belongs to* $\mathscr{C}(\Delta_I \times M)$ *and we have*

$$\|\Phi\| = \varepsilon_U[x_i, I] .$$

*Proof.* Since for each positive number $\delta$ there is a set $\mathfrak{i}_0 \in \mathfrak{F}(I)$ such that

$$\sum_{I \setminus \mathfrak{i}_0} |\langle x_i, a \rangle| \leqq \delta \quad \text{for} \quad a \in M \subset U^0 ,$$

the functions

$$\Phi_{\mathfrak{i}}(\alpha_i, a) = \sum_{\mathfrak{i}} \alpha_i \langle x_i, a \rangle$$

satisfy the inequality

$$|\Phi(\alpha_i, a) - \Phi_{\mathfrak{i}}(\alpha_i, a)| \leqq \delta \quad \text{for all} \quad \mathfrak{i} \in \mathfrak{F}(I) \text{ with } \mathfrak{i} \supset \mathfrak{i}_0 .$$

Consequently the continuous functions $\Phi_{\mathfrak{i}}$ on $\Delta_I \times M$ converge uniformly to $\Phi$. Hence the function $\Phi$ is also continuous.

Since the relation

$$|\Phi(\alpha_i, a)| \leqq \sum_I |\langle x_i, a \rangle| \leqq \varepsilon_U[x_i, I]$$

holds for $[\alpha_i, I] \in \Delta_I$ and $a \in M$, we have

$$\|\Phi\| \leqq \varepsilon_U[x_i, I] .$$

Conversely, if we determine for each linear form $a \in U^0$ real or complex numbers $\alpha_i$ with $|\langle x_i, a \rangle| = \alpha_i \langle x_i, a \rangle$ and $|\alpha_i| = 1$, then for all sets $\mathfrak{i} \in \mathfrak{F}(I)$ we obtain the relation

$$\sum_{\mathfrak{i}} |\langle x_i, a \rangle| = \left\langle \sum_{\mathfrak{i}} \alpha_i x_i, a \right\rangle \leqq p_U\left(\sum_{\mathfrak{i}} \alpha_i x_i\right)$$
$$\leqq \sup\left\{\left|\left\langle \sum_{\mathfrak{i}} \alpha_i x_i, b \right\rangle\right| : b \in M\right\} \leqq \|\Phi_{\mathfrak{i}}\| ,$$

since

$$\lim_i \|\Phi_i\| = \|\Phi\|$$

we obtain

$$\sum_I |\langle x_i, a\rangle| \leqq \|\Phi\| \quad \text{for all} \quad a \in U^0$$

by passing to the limit. Thus the assertion

$$\varepsilon_U[x_i, I] \leqq \|\Phi\|$$

is proved so that the required equality is also true.

**2.3.3.** If $E$ and $F$ are two normed spaces with closed unit balls $U$ and $V$, then for an arbitrary essential subset $M$ of $E'$ we have the

**Theorem.** *A mapping $T \in \mathfrak{L}(E, F)$ is absolutely summing if and only if there is on $M$ a positive Radon measure $\mu$, such that the relation*

$$p_V(T\,x) \leqq \int_M |\langle x, a\rangle|\, d\mu \quad \text{for} \quad x \in E$$

*is valid. If $\pi_M(T)$ is the collection of all such Radon measures we then have*

$$\boldsymbol{\pi}(T) = \inf\{\mu(M) : \mu \in \pi_M(T)\}\ ,$$

*and there is in $\pi_M(T)$ at least one mesaure $\mu_0$ with $\boldsymbol{\pi}(T) = \mu_0(M)$.*

*Proof.* We determine an index set $I$ so that a map $i \to b_i$ exists from $I$ onto $V^0$. Then if $T$ is an absolutely summing mapping from $E$ into $F$, a continuous linear form $\boldsymbol{a}$ is defined on $\boldsymbol{l}_I^1(E)$ by the equation

$$\langle [x_i, I], \boldsymbol{a}\rangle = \sum_I \langle T\,x_i, b_i\rangle \tag{1}$$

because we have

$$|\langle [x_i, I], \boldsymbol{a}\rangle| \leqq \sum_I |\langle T\,x_i, b_i\rangle| \leqq \sum_I p_V(T\,x_i) \leqq \boldsymbol{\pi}_V[T\,x_i, I] \leqq \boldsymbol{\pi}(T)\,\varepsilon_U[x_i, I]\,.$$

By the previous lemma, $\boldsymbol{l}_I^1(E)$ can be considered as a linear subspace of $\mathscr{C}(\Delta_I \times M)$ by identifying the families $[x_i, I]$ with the continuous function $\Phi$. Thus by the Hahn Banach Theorem 0.4.4 there is a linear form $M_0$ on $\mathscr{C}(\Delta_I \times M)$ (i. e. a Radon measure) with

$$\langle [x_i, I], \boldsymbol{a}\rangle = \int_{\Delta_I \times M} \Phi(\alpha_i, a)\, dM_0 \quad \text{for} \quad [x_i, I] \in \boldsymbol{l}_I^1(E)\ , \tag{2}$$

whose norm is no larger than $\boldsymbol{\pi}(T)$.

Now we obtain the desired positive Radon measure $\mu_0$ from the equation

$$\langle \varphi, \mu_0\rangle = \int_M \varphi(a)\, d\mu_0 = \int_{\Delta_I \times M} \varphi(a)\, d|M_0| \quad \text{for} \quad \varphi \in \mathscr{C}(M)\,. \tag{3}$$

Here it is true that

$$\mu_0(M) \leqq \boldsymbol{\pi}(T)\,. \tag{4}$$

If we substitute in (2) the family $[x\,\delta_i^{[j]}, I]$ with $x \in E$, we obtain the identity

$$\langle T\,x, b_j\rangle = \int\limits_{\Delta_I \times M} \alpha_j \langle x, a\rangle \, dM_0 .$$

This leads to the inequality

$$|\langle T\,x, b_j\rangle| \leqq \int\limits_{\Delta_I \times M} |\langle x, a\rangle| \, d|M_0| = \int\limits_M |\langle x, a\rangle| \, d\mu_0 .$$

By assumption, for each linear form $b \in V^0$ there is an index $j \in I$ with $b = b_j$, so that we have

$$|\langle T\,x, b\rangle| \leqq \int\limits_M |\langle x, a\rangle| \, d\mu_0 \quad \text{for all} \quad b \in V^0 .$$

This establishes the inequality

$$p_V(T\,x) = \sup\{|\langle T\,x, b\rangle| : b \in V^0\} \leqq \int\limits_M |\langle x, a\rangle| \, d\mu_0 .$$

On the other hand, suppose for a mapping $T \in \mathscr{L}(E, F)$ on $M$ there is a positive Radon measure $\mu$ with

$$p_V(T\,x) \leqq \int\limits_M |\langle x, a\rangle| \, d\mu \quad \text{for} \quad x \in E .$$

Then for each finite family $[x_n, \mathfrak{n}]$ from $E$ we have the inequality

$$\sum_{\mathfrak{n}} p_V(T\,x_n) \leqq \int\limits_M \sum_{\mathfrak{n}} |\langle x_n, a\rangle| \, d\mu \leqq \mu(M) \sup\left\{\sum_{\mathfrak{n}} |\langle x_n, a\rangle| : a \in U^0\right\} .$$

Consequently, the mapping $T$ is absolutely summing by 2.2.1. Moreover, we have

$$\boldsymbol{\pi}(T) \leqq \mu(M) ,$$

which leads to the inequality

$$\boldsymbol{\pi}(T) \leqq \inf\{\mu(M) : \mu \in \pi_M(T)\} . \tag{5}$$

However, since relation (4) holds for the Radon measure $\mu_0$ constructed at the beginning of our proof, we must, in fact, have equality in (4).

**2.3.4.** For $M$ an arbitrary compact Hausdorff space we have the following special case of the preceding theorem.

**Proposition.** *A continuous linear mapping $T$ from $\mathscr{C}(M)$ into a normed space $F$ is absolutely summing if and only if there is a positive Radon measure $\mu$ on $M$ for which the relation*

$$p_V(T\,\varphi) \leqq \int\limits_M |\varphi(x)| \, d\mu \quad \textit{for all} \quad \varphi \in \mathscr{C}(M)$$

*is valid.*

*Proof.* Our assertion follows immediately from the fact, established in 2.3.1, that $M$ can be identified with the essential set of all Dirac measures $\delta_x$. It should be noted that the weak topology induced on $M$ by $U^0$ coincides with the original topology because of 0.1.9.

## 2.4. A Special Absolutely Summing Mapping

**2.4.1.** For the following discussion we need an inequality from the theory of orthogonal series. For this purpose, we consider the **Rademacher functions** which are defined on the interval $[0, 1]$ for $p = 1, 2, \ldots$ by the equation

$$\varrho_p(t) = \begin{cases} +1 \text{ for } k/2^p < t < (k+1)/2^p & \text{with } k = 0, 2, \ldots, 2^p - 2 \\ \phantom{+}0 \text{ for } t = k/2^p & \text{with } k = 0, 1, \ldots, 2^p \\ -1 \text{ for } k/2^p < t < (k+1)/2^p & \text{with } k = 1, 3, \ldots, 2^p - 1 \,. \end{cases}$$

The Rademacher functions form an orthogonal family because of the relation

$$\int_0^1 \varrho_p(t)\, \varrho_q(t)\, dt = \delta_{pq}\,.$$

To see this, note first that when $p = q$ the product $\varrho_p(t)\, \varrho_q(t)$ is equal to 1 except at finitely many points. On the other hand, let the two natural numbers $p$ and $q$ be different and assume, without loss of generality, that $q$ is larger than $q$. Since the function $\varrho_p(t)$ is, by definition, equal to $(-1)^k$ on the interval

$$\Delta_p^k = (k/2^p, (k+1)/2^p)$$

we have

$$\int_0^1 \varrho_p(t)\, \varrho_q(t)\, dt = \sum_{k=0}^{2^p-1} (-1)^k \int_{\Delta_p^k} \varrho_q(t)\, dt = 0\,.$$

It is proved in the same way that the integral

$$\int_0^1 \varrho_p(t)\, \varrho_q(t)\, \varrho_r(t)\, \varrho_s(t)\, dt$$

has the value 1 or 0 according to whether $p$, $q$, $r$ and $s$ are pairwise equal or not.

For real or complex numbers $\xi_1, \ldots, \xi_h$ we set

$$\alpha = \sum_1^h |\xi_p|^2 = \int_0^1 \left| \sum_1^h \xi_p\, \varrho_p(t) \right|^2 dt\,.$$

Then, since

$$\int_0^1 \left| \sum_1^h \xi_p\, \varrho_p(t) \right|^4 dt = \sum_{p=1}^h \sum_{q=1}^h \sum_{r=1}^h \sum_{s=1}^h \xi_p\, \xi_q\, \bar{\xi}_r\, \bar{\xi}_s \int_0^1 \varrho_p(t)\, \varrho_q(t)\, \varrho_r(t)\, \varrho_s(t)\, dt$$

the inequality

$$\int_0^1 \left| \sum_1^h \xi_p \varrho_p(t) \right|^4 dt \leqq 3\alpha^2$$

is valid. From the Hölder inequality with the conjugate exponents $3/2$ and $3$ we obtain the relation

$$\alpha = \int_0^1 \left| \sum_1^h \xi_p \varrho_p(t) \right|^{2/3} \left| \sum_1^h \xi_p \varrho_p(t) \right|^{4/3} dt \leqq \left\{ \int_0^1 \left| \sum_1^h \xi_p \varrho_p(t) \right| dt \right\}^{2/3} \{3\alpha^2\}^{1/3}$$

which yields the **Khinchin inequality**

$$\left\{ \sum_1^h |\xi_p|^2 \right\}^{1/2} \leqq \sqrt{3} \int_0^1 \left| \sum_1^h \xi_p \varrho_p(t) \right| dt .$$

by an easy transformation.

**2.4.2.** We now consider for an arbitrary index set $I$ the Banach space $\boldsymbol{l}_I^1$ as well as the Hilbert space $\boldsymbol{l}_I^2$ of quadratically summable families of numbers $[\xi_i, I]$ with the norm

$$\boldsymbol{\lambda}_2[\xi_i, I] = \left\{ \sum_I |\xi_i|^2 \right\}^{1/2} .$$

We then have the

**Proposition.** *The identity mapping $R'$ from $\boldsymbol{l}_I^1$ into $\boldsymbol{l}_I^2$ is absolutely summing and $\boldsymbol{\pi}(R') \leqq \sqrt{3}$.*

**Proof.** Let $\mathfrak{i} = \{i_1, \ldots, i_h\}$ be an arbitrary set from $\mathfrak{F}(I)$; then the equation

$$\langle [\xi_i, I], \varrho(t) \rangle = \sum_1^h \xi_{i_p} \varrho_p(t) \quad \text{for } 0 \leqq t \leqq 1$$

defines a continouus linear form $\varrho(t)$ on $\boldsymbol{l}_I^1$, which lies in the closed unit ball of $(\boldsymbol{l}_I^1)'$.

We now consider a finite family $[x_n, \mathfrak{n}]$ from $\boldsymbol{l}_I^1$ with $x_n = [\xi_i^{[n]}, I]$. Then for all $n \in \mathfrak{n}$, the inequality

$$\left\{ \sum_{\mathfrak{i}} |\xi_i^{[n]}|^2 \right\}^{1/2} \leqq \sqrt{3} \int_0^1 |\langle x_n, \varrho(t) \rangle| \, dt$$

is valid because of 2.4.1. If the closed unit ball of $\boldsymbol{l}_I^1$ is denoted by $U$, we obtain by summation over $n$ the relation

$$\sum_{\mathfrak{n}} \left\{ \sum_{\mathfrak{i}} |\xi_i^{[n]}|^2 \right\}^{1/2} \leqq \sqrt{3} \int_0^1 \sum_{\mathfrak{n}} |\langle x_n, \varrho(t) \rangle| \, dt \leqq \sqrt{3} \sup \left\{ \sum_{\mathfrak{n}} |\langle x_n, a \rangle| : a \in U^0 \right\},$$

which is true because $\varrho(t) \in U^0$. Since this relation holds for all sets $\mathfrak{i} \in \mathfrak{F}(I)$, we have

$$\sum_{\mathfrak{n}} \boldsymbol{\lambda}_2(R' x_n) \leqq \sqrt{3} \sup \left\{ \sum_{\mathfrak{n}} |\langle x_n, a \rangle| : a \in U^0 \right\} .$$

Therefore, because of Proposition 2.2.1, the mapping $R'$ is absolutely summing and $\boldsymbol{\pi}(R') \leqq \sqrt{3}$ holds.

**2.4.3.** Since two distinct families $e_j = [\delta_i^{[j]}, I]$ in $\boldsymbol{l}_I^2$ have distance $\boldsymbol{\lambda}_2[e_j - e_k] = \sqrt{2}$ from one another, the range of $R'$ for an uncountable index set $I$ cannot be separable. Consequently, we have the

**Proposition.** *There exist absolutely summing mappings without separable range.*

**2.4.4.** Since the closed unit ball of $\boldsymbol{l}_I^1$ for an infinite index set $I$ is not a precompact subset of $\boldsymbol{l}_I^2$, we obtain the

**Proposition.** *There exist absolutely summing mappings which are not precompact.*

**2.4.5.** As an immediate consequence of 2.4.4 we obtain finally the

**Proposition.** *There exist absolutely summing mappings which cannot be approximated with respect to the $\boldsymbol{\pi}$-norm by finite mappings.*

**2.4.6.** If we denote the identity mapping from $\boldsymbol{l}_N^2$ into $\boldsymbol{c}_N$ by $R$, $R'$ resp. $R''$ is the identity mapping from $\boldsymbol{l}_N^1$ into $\boldsymbol{l}_N^2$ resp. from $\boldsymbol{l}_N^2$ into $\boldsymbol{m}_N$.

We now show that the mappings $R$ and $R''$ are not absolutely summing. For this purpose we consider the families of numbers $x_m = [1/m\, \delta_n^{[m]}, N]$. Since each linear form $a \in \boldsymbol{l}_N^2$ can be written in the form

$$\langle [\xi_n, N], a \rangle = \sum_N \alpha_n \xi_n \quad \text{with} \quad [\alpha_n, N] \in \boldsymbol{l}_N^2,$$

the inequality

$$\sum_M |\langle x_m, a \rangle| = \sum_M (1/m)|\alpha_m| \leqq \left\{\sum_M (1/m^2)\right\}^{1/2} \left\{\sum_M |\alpha_m|^2\right\}^{1/2} < +\infty$$

holds for $M = \{1, 2, \ldots\}$. Consequently, the family $[x_m, M]$ in $\boldsymbol{l}_N^2$ is weakly summable. If the mapping $R$, resp. $R''$, were absolutely summing, the family $[R\, x_m, M]$, resp. $[R''\, x_m, M]$, in $\boldsymbol{c}_N$, resp. $\boldsymbol{m}_N$, would have to be absolutely summable. But this is impossible because

$$\sum_M \boldsymbol{\lambda}_\infty[R\, x_m] = \sum_M \boldsymbol{\lambda}_\infty[R''\, x_m] = \sum_M 1/m = +\infty .$$

Taking into account the fact proved in Proposition 2.4.2 that the mapping $R'$ is absolutely summing we obtain the

**Proposition.** *There are absolutely summing (non-absolutely summing) mappings which have non-absolutely summing (absolutely summing) dual mappings.*

## 2.5. Hilbert-Schmidt-Mappings

**2.5.1.** Let $E$ and $F$ be two arbitrary real or complex Hilbert spaces with closed unit balls $U$ and $V$. A mapping $T \in \mathscr{L}(E, F)$ is called a **Hilbert-Schmidt-mapping**, if for two complete orthonormal systems $[e_i, I]$ and $[f_j, J]$ from $E$, resp. $F$, the inequality

$$\sigma(T)^2 = \sum_{I,J} |(T\, e_i, f_j)|^2 < +\infty$$

holds. Here $\sigma(T)$ is independent of the choice of the two complete orthonormal systems $[e_i, I]$ and $[f_j, J]$ because

$$\sigma(T)^2 = \sum_{I,J} |(T\, e_i, f_j)|^2 = \sum_I p_V(T\, e_i)^2$$

and

$$\sigma(T)^2 = \sum_{I,J} |(e_i, T^* f_j)|^2 = \sum_J p_U(T^* f_j)^2 .$$

**2.5.2.** If $\mathscr{S}(E, F)$ is the collection of all Hilbert-Schmidt-mappings from $E$ into $F$ we have the

**Proposition.** *$\mathscr{S}(E, F)$ is a linear space with the scalar product*

$$(S, T) = \sum_I (S\, e_i, T\, e_i) .$$

*Proof.* For each mapping $T \in \mathscr{S}(E, F)$, the relation

$$p_V(T\, x) = p_V\Big(\sum_I (x, e_i)\, T\, e_i\Big) \leqq \sum_I |(x, e_i)|\, p_V(T\, e_i)$$
$$\leqq \Big\{\sum_I |(x, e_i)|^2\Big\}^{1/2} \Big\{\sum_I p_V(T\, e_i)^2\Big\}^{1/2} = p_U(x)\, \sigma(T)$$

implies the inequality

$$\beta(T) \leqq \sigma(T) . \tag{0}$$

Since for two mappings $S, T \in \mathscr{S}(E, F)$ we have the inequality

$$\Big\{\sum_I p_V(S\, e_i + T\, e_i)^2\Big\}^{1/2} \leqq \Big\{\sum_I p_V(S\, e_i)^2\Big\}^{1/2} + \Big\{\sum_I p_V(T\, e_i)^2\Big\}^{1/2},$$

it follows that $S + T$ is a Hilbert-Schmidt-mapping and the triangle inequality

$$\sigma\,(S + T) \leqq \sigma(S) + \sigma(T) \tag{1}$$

is valid.

Since it is easily seen that for $T \in \mathscr{S}(E, F)$ and each number $\alpha$ the mapping $\alpha\, T$ also belongs to $\mathscr{S}(E, F)$ and since

$$\sigma(\alpha\, T) = |\alpha|\, \sigma(T) , \tag{2}$$

it follows that $\mathscr{S}(E, F)$ is a linear space with the norm $\sigma(T)$.

Finally since

$$\sum_I |(S\,e_i,\,T\,e_i)| \leqq \left\{\sum_I p_V(S\,e_i)^2\right\}^{1/2} \left\{\sum_I p_V(T\,e_i)^2\right\}^{1/2},$$

the family of numbers $[(S\,e_i,\,T\,e_i),\,I]$ is summable, and we have the expression

$$(S,\,T) = \sum_I (S\,e_i,\,T\,e_i) \quad \text{for all} \quad S,\,T \in \mathcal{S}(E,\,F)\,.$$

From this equation it immediately follows that $(S,\,T)$ is a scalar product on $\mathcal{S}(E,\,F)$ for which $\boldsymbol{\sigma}(T) = (T,\,T)^{1/2}$.

**2.5.3. Proposition.** *The linear space $\mathcal{A}(E,\,F)$ is dense in $\mathcal{S}(E,\,F)$.*

*Proof.* For each mapping $T \in \mathcal{A}(E,\,F)$ we can determine a complete orthonormal system $[e_i,\,I]$ in $E$ which contains only finitely many elements $e_i$ with $T\,e_i \neq$ o. Consequently, we have

$$\boldsymbol{\sigma}(T)^2 = \sum_I p_V(T\,e_i)^2 < +\infty\,,$$

and the mapping $T$ belongs to $\mathcal{S}(E,\,F)$. The relation

$$\mathcal{A}(E,\,F) \subset \mathcal{S}(E,\,F)$$

is thus proved.

For $T$ an arbitrary mapping from $\mathcal{S}(E,\,F)$, we consider the mappings

$$T_{\mathfrak{i}}(x) = \sum_{\mathfrak{i}} (x,\,e_i)\,T\,e_i \qquad \text{with} \quad \mathfrak{i} \in \mathfrak{F}(I)$$

for an arbitrary complete orthonormal system $[e_i,\,I]$ in $E$. These mappings are all finite.

Now because

$$\sum_I p_V(T\,e_i)^2 < +\infty$$

there is, for each positive number $\delta$, a set $\mathfrak{i}_0 \in \mathfrak{F}(I)$ with

$$\sum_{I \setminus \mathfrak{i}_0} p_V(T\,e_i)^2 \leqq \delta^2;$$

therefore, for all $\mathfrak{i} \in \mathfrak{F}(I)$ with $\mathfrak{i} \supset \mathfrak{i}_0$ the inequality

$$\boldsymbol{\sigma}\,(T - T_{\mathfrak{i}})^2 \leqq \sum_{I \setminus \mathfrak{i}} p_V(T\,e_i)^2 \leqq \delta^2$$

holds. Consequently, we have

$$\boldsymbol{\sigma} - \lim_{\mathfrak{i}} T_{\mathfrak{i}} = T\,.$$

**2.5.4. Proposition.** *Every Hilbert-Schmidt-mapping is compact.*

*Proof.* For $T$ an arbitrary mapping from $\mathcal{S}(E,\,F)$, the mappings $T_{\mathfrak{i}}$ constructed in 2.5.3 converge in $\mathcal{L}(E,\,F)$ to $T$ because

$$\boldsymbol{\beta}\,(T - T_{\mathfrak{i}}) \leqq \boldsymbol{\sigma}\,(T - T_{\mathfrak{i}})\,.$$

Consequently, $T$ is compact by 0.10.6.

**2.5.5.** We now present an interesting characterization of Hilbert-Schmidt-mappings.

**Theorem.** *For two Hilbert spaces E and F the absolutely summing mappings coincide with the Hilbert-Schmidt-mappings from E into F, and for each mapping* $T \in \mathcal{S}(E, F) = \mathfrak{P}(E, F)$ *we have*

$$\boldsymbol{\sigma}(T) \leqq \boldsymbol{\pi}(T) \leqq \sqrt{3}\, \boldsymbol{\sigma}(T) .$$

*Proof.* Since each mapping $T \in \mathcal{S}(E, F)$ is compact we have by the Spectral Decomposition Theorem 8.2.1 two orthonormal systems $[e_i, I]$ and $[f_i, I]$ in $E$, resp. $F$, as well as a family of numbers $[\lambda_i, I] \in \boldsymbol{c}_I$ such that

$$T x = \sum_I \lambda_i (x, e_i) f_i \quad \text{for } x \in E .$$

Since $(T e_i, f_i) = \lambda_i$, the inequality

$$\sum_I |\lambda_i|^2 \leqq \boldsymbol{\sigma}(T)^2 < +\infty$$

holds.

We now define two mappings $T_1 \in \mathfrak{L}(E, \boldsymbol{l}_I^1)$ and $T_2 \in \mathfrak{L}(\boldsymbol{l}_I^2, F)$ by the equations

$$T_1 x = [\lambda_i (x, e_i), I] \quad \text{for } x \in E$$

and

$$T_2 [\xi_i, I] = \sum_I \xi_i f_i \quad \text{for } [\xi_i, I] \in \boldsymbol{l}_I^2 .$$

For these two mappings, we have the relations

$$\boldsymbol{\beta}(T_1) \leqq \boldsymbol{\sigma}(T) \quad \text{and} \quad \boldsymbol{\beta}(T_2) \leqq 1$$

because

$$\sum_I [\lambda_i (x, e_i)] \leqq \boldsymbol{\sigma}(T)\, p_U(x) \quad \text{and} \quad p_V\left(\sum_I \xi_i f_i\right) \leqq \boldsymbol{\lambda}_2[\xi_i, I] .$$

Finally if $R'$ is the identity mapping from $\boldsymbol{l}_I^1$ into $\boldsymbol{l}_I^2$ with $\boldsymbol{\pi}(R') \leqq \sqrt{3}$ we have $T = T_2 R' T_1$. Consequently, $T$ is absolutely summing by Proposition 2.2.5 and we have $\boldsymbol{\pi}(T) \leqq \sqrt{3}\, \boldsymbol{\sigma}(T)$.

Conversely, for each absolutely summing mapping $T \in \mathfrak{P}(E, F)$ we have the inequality

$$\boldsymbol{\pi}_V[T x_i, I] \leqq \boldsymbol{\pi}(T)\, \boldsymbol{\varepsilon}_U[x_i, I] \quad \text{for all } [x_i, I] \in \boldsymbol{l}_I^1[E]$$

for each index set $I$. If $[e_i, I]$ is a complete orthonormal system in $E$, then the families $[\xi_i e_i, I]$ with $[\xi_i, I] \in \boldsymbol{l}_I^2$ belong to $\boldsymbol{l}_I^1[E]$ because the relation

$$\sum_I |\langle \xi_i e_i, a\rangle| \leqq \left\{\sum_I |\xi_i|^2\right\}^{1/2} \left\{\sum_I |\langle e_i, a\rangle|^2\right\}^{1/2} \leqq \boldsymbol{\lambda}_2[\xi_i, I]$$

holds for all linear forms $a \in U^0$ by the Bessel inequality 0.9.4. Thus the relation

$$\boldsymbol{\varepsilon}_U [\xi_i e_i, I] \leqq \boldsymbol{\lambda}_2[\xi_i, I] \quad \text{for all} \quad [\xi_i, I] \in \boldsymbol{l}_I^2$$

is also proved and from this we get the inequality

$$\sum_I |\xi_i| \, p_V(T e_i) = \boldsymbol{\pi}_V [T(\xi_i e_i), I] \leqq \boldsymbol{\pi}(T) \, \boldsymbol{\varepsilon}_U[\xi_i e_i, I] \leqq \boldsymbol{\pi}(T) \, \boldsymbol{\lambda}_2[\xi_i, I] .$$

Therefore by 1.1.7 we have

$$\boldsymbol{\sigma}(T) = \left\{\sum_I p_V(T e_i)^2\right\}^{1/2} \leqq \boldsymbol{\pi}(T) < +\infty .$$

**2.5.6.** It can be shown (see A. Grothendieck [5], p. 55), that in real Hilbert space the relation

$$\boldsymbol{\pi}(T) \leqq \sqrt{\pi/2} \, \boldsymbol{\sigma}(T)$$

holds, and that it cannot be improved.

Chapter 3

# Nuclear Mappings

*Nuclear mappings* were first introduced under the name "operators of the trace-class" when R. Schatten and J. Von Neumann investigated the question of which continuous linear mappings of a Hilbert space determine a meaningful trace. The extension of these ideas to Banach spaces led A. Grothendieck [3] to the concept of nuclear mapping. However, the original object was relegated to the background, and it is still unknown whether every nuclear mapping from a Banach space into itself has a uniquely defined trace.

In this chapter we shall consider nuclear mappings from a normed space $E$ into a normed space $F$. It will be shown that the collection of these mappings forms a linear space on which a suitable norm can be introduced.

A. Grothendieck ([3], Chap. I, p. 88) has noted the following very undesirable property possessed by nuclear mappings. If $F$ is embedded into a larger normed space $G$ it is possible that the mapping $T \in \mathcal{L}(E, F)$ is nuclear as a mapping from $E$ into $G$ but not as a mapping from $E$ into $F$. The detailed investigation of these matters led the present author to the concept of a *quasinuclear mapping* (Proposition 3.2.7).

The most important result of this chapter is Theorem 3.3.5, due to A. Grothendieck ([3], Chap. I, p. 162), which we shall prove in a new and very simple way. It consists of the profound assertion that *the product of two absolutely summing mappings is always nuclear.*

The first application of the result thus obtained is a proof of the theorem of Dvoretzky and Rogers which appears at the end of the chapter.

## 3.1. Nuclear Mappings in Normed Spaces

**3.1.1.** Let $E$ and $F$ be two arbitrary normed spaces with closed unit balls $U$ and $V$. A mapping $T \in \mathcal{L}(E, F)$ is called **nuclear** if there are continuous linear forms $a_n \in E'$ and elements $y_n \in F$ with

(1)
$$\sum_N p_{U^0}(a_n)\, p_V(y_n) < +\infty$$

such that $T$ has the form

$$T\,x = \sum_N \langle x, a_n\rangle\, y_n \quad \text{for} \quad x \in E\,.$$

For each nuclear mapping $T$ we set

$$\boldsymbol{\nu}(T) = \inf\left\{\sum_N p_{U^\circ}(a_n)\, p_V(y_n)\right\},$$

where the infimum is taken over all possible representations of $T$.

**3.1.2.** If the set of all nuclear mappings from $E$ into $F$ is designated $\mathcal{N}(E, F)$, we then have the

**Proposition.** *$\mathcal{N}(E, F)$ is a linear space with the norm $\boldsymbol{\nu}(T)$.*

*Proof.* If $T$ is an arbitrary mapping from $\mathcal{N}(E, F)$, then for each representation of $T$ of the form

$$T\,x = \sum_N \langle x, a_n\rangle\, y_n\,,$$

we have

$$\boldsymbol{\beta}(T) \leqq \sum_N p_{U^\circ}(a_n)\, p_V(y_n)$$

because

$$p_V(T\,x) \leqq p_U(x) \sum_N p_{U^\circ}(a_n)\, p_V(y_n)\,.$$

Consequently,

$$\boldsymbol{\beta}(T) \leqq \boldsymbol{\nu}(T)\,. \tag{0}$$

Suppose for $S$, $T \in \mathcal{N}(E, F)$ and arbitrary $\delta > 0$ we have the two representations

$$S\,x = \sum_N \langle x, a_n\rangle\, y_n \quad \text{and} \quad T\,x = \sum_N \langle x, b_n\rangle\, z_n$$

with

$$\sum_N p_{U^\circ}(a_n)\, p_V(y_n) \leqq \boldsymbol{\nu}(S) + \delta \quad \text{and} \quad \sum_N p_{U^\circ}(b_n)\, p_V(z_n) \leqq \boldsymbol{\nu}(T) + \delta\,.$$

The mapping $S + T$ then has the form

$$(S + T)\,x = \sum_N \langle x, a_n\rangle\, y_n + \sum_N \langle x, b_n\rangle\, z_n$$

and we have

$$\boldsymbol{\nu}(S + T) \leqq \sum_N p_{U^\circ}(a_n)\, p_V(y_n) + \sum_N p_{U^\circ}(b_n)\, p_V(z_n) \leqq \boldsymbol{\nu}(S) + \boldsymbol{\nu}(T) + 2\delta\,.$$

Therefore, the mapping $S + T$ belongs to $\mathcal{N}(E, F)$ and the triangle inequality

$$\boldsymbol{\nu}(S + T) \leqq \boldsymbol{\nu}(S) + \boldsymbol{\nu}(T) \tag{1}$$

is valid.

As one can easily see, for $T \in \mathcal{N}(E, F)$ and each number $\alpha$, $\alpha T$ is also nuclear and

$$\boldsymbol{\nu}(\alpha T) = |\alpha|\, \boldsymbol{\nu}(T)\,, \tag{2}$$

so that our assertion is completely proved.

**3.1.3. Lemma.** *If $\{T_\alpha\}$ is a directed Cauchy-system from $\mathcal{N}(E, F)$ and if there is a mapping $T \in \mathcal{L}(E, F)$ with $\lim\limits_\alpha T_\alpha x = T x$ for $x \in E$, then $T$ belongs to $\mathcal{N}(E, F)$ and $\boldsymbol{\nu} - \lim\limits_\alpha T_\alpha = T$.*

*Proof.* We determine a monotonically increasing sequence of indices $\alpha_k$ for which

$$\boldsymbol{\nu}\,(T_\alpha - T_\beta) < 1/2^{k+2} \quad \text{for } \alpha, \beta \geqq \alpha_k\,.$$

Then, the nuclear mappings $T_{\alpha_{k+1}} - T_{\alpha_k}$ for $k = 1, 2, \ldots$ can be written

$$(T_{\alpha_{k+1}} - T_{\alpha_k})\, x = \sum_N \langle x, a_n^{[k]}\rangle\, y_n^{[k]}$$

with

$$\sum_N p_{U^\circ}(a_n^{[k]})\, p_V(y_n^{[k]}) < 1/2^{k+2}\,.$$

Consequently, for $p = 1, 2, \ldots$, we have

$$(T_{\alpha_{k+p}} - T_{\alpha_k})\, x = \sum_{h=k}^{k+p-1} \sum_N \langle x, a_n^{[h]}\rangle\, y_n^{[h]}\,,$$

and by taking the limit as $p \to \infty$, we obtain the identity

$$(T - T_{\alpha_k})\, x = \sum_{h=k}^{\infty} \sum_N \langle x, a_n^{[h]}\rangle\, y_n^{[h]}\,.$$

Since the inequality

$$\boldsymbol{\nu}\,(T - T_{\alpha_k}) \leqq \sum_{h=k}^{\infty} \sum_N p_{U^\circ}(a_n^{[h]})\, p_V(y_n^{[h]}) \leqq 1/2^{k+1}$$

holds, the mapping $T - T_{\alpha_k}$ is nuclear and hence so is $T$. Finally, since the inequality

$$\boldsymbol{\nu}\,(T - T_\alpha) \leqq \boldsymbol{\nu}\,(T - T_{\alpha_k}) + \boldsymbol{\nu}\,(T_{\alpha_k} - T_\alpha) \leqq 1/2^k \quad \text{for } \alpha \geqq \alpha_k$$

is valid, we have the assertion

$$\boldsymbol{\nu} - \lim_\alpha T_\alpha = T\,.$$

As a simple consequence of the lemma just proved we have the

**Proposition.** *For a normed space $E$ and a Banach space $F$, $\mathcal{N}(E, F)$ is a Banach space.*

**3.1.4. Proposition.** *The linear space $\mathcal{A}(E, F)$ is dense in $\mathcal{N}(E, F)$.*

*Proof.* Since each finite mapping $T$ can be expressed in the form

$$T x = \sum_{n=1}^{k} \langle x, a_n \rangle y_n \quad \text{with} \quad a_1, \ldots, a_k \in E' \quad \text{and} \quad y_1, \ldots, y_k \in F ,$$

$\mathcal{A}(E, F)$ is a linear subspace of $\mathcal{N}(E, F)$.

We may consider an arbitrary nuclear mapping $T \in \mathcal{N}(E, F)$. There are then continuous linear forms $a_n \in E'$ and elements $y_n \in F$ with

$$T x = \sum_{N} \langle x, a_n \rangle y_n \quad \text{for} \quad x \in E$$

and

$$\sum_{N} p_{U^0}(a_n)\, p_V(y_n) < +\infty .$$

Since for each positive number $\delta$ we can find a set $\mathfrak{n}_0 \in \mathfrak{F}(N)$ with

$$\sum_{N \setminus \mathfrak{n}_0} p_{U^0}(a_n)\, p_V(y_n) \leqq \delta ,$$

we have for the finite mappings

$$T_{\mathfrak{n}}\, x = \sum_{\mathfrak{n}} \langle x, a_n \rangle y_n$$

the inequality

$$\boldsymbol{\nu}\,(T - T_{\mathfrak{n}}) \leqq \delta \qquad \text{for all} \quad \mathfrak{n} \in \mathfrak{F}(N) \quad \text{with} \quad \mathfrak{n} \supset \mathfrak{n}_0 .$$

Thus

$$\boldsymbol{\nu} - \lim_{\mathfrak{n}} T_{\mathfrak{n}} = T .$$

**3.1.5. Proposition.** *Each nuclear mapping is precompact.*

*Proof.* If $T$ is an arbitrary mapping from $\mathcal{N}(E, F)$, then the mappings $T_{\mathfrak{n}}$ constructed in 3.1.4 converge in $\mathcal{L}(E, F)$ to $T$ since

$$\boldsymbol{\beta}\,(T - T_{\mathfrak{n}}) \leqq \boldsymbol{\nu}\,(T - T_{\mathfrak{n}}) .$$

Consequently, $T$ is precompact.

**3.1.6.** As an immediate consequence of 3.1.5 and 0.10.6 we obtain the

**Proposition.** *Each nuclear mapping has a separable range.*

**3.1.7.** Let $E$, $F$ and $G$ be three normed spaces with closed unit balls $U$, $V$ and $W$. We then obtain the

**Proposition.**

(1) *If $T \in \mathcal{L}(E, F)$ and $S \in \mathcal{N}(F, G)$, then $ST \in \mathcal{N}(E, G)$ and* $\boldsymbol{\nu}(ST) \leqq \boldsymbol{\nu}(S)\, \boldsymbol{\beta}(T)$.

(2) *If $T \in \mathcal{N}(E, F)$ and $S \in \mathcal{L}(F, G)$, then $ST \in \mathcal{N}(E, G)$ and* $\boldsymbol{\nu}(ST) \leqq \boldsymbol{\beta}(S)\, \boldsymbol{\nu}(T)$.

*Proof.*

(1) Since the mapping $S$ is nuclear, for each positive number $\delta$ there are continuous linear forms $b_n \in F'$ and elements $z_n \in G$ with

$$S y = \sum_N \langle y, b_n \rangle z_n \quad \text{for } y \in F$$

and

$$\sum_N p_{V^\circ}(b_n)\, p_W(z_n) \leqq \boldsymbol{\nu}(S) + \delta .$$

Consequently, the mapping $ST$ has the form

$$S T x = \sum_N \langle x, T' b_n \rangle z_n \quad \text{for } x \in E ,$$

and

$$\begin{aligned} \boldsymbol{\nu}(S T) &\leqq \sum_N p_{U^\circ}(T' b_n) p_W(Z_n) \leqq \boldsymbol{\beta}(T') \sum_N p_{V^\circ}(b_n)\, p_W(z_n) \\ &\leqq [\boldsymbol{\nu}(S) + \delta]\, \boldsymbol{\beta}(T) . \end{aligned}$$

Therefore, the operator $S\,T$ is nuclear and

$$\boldsymbol{\nu}(S\,T) \leqq \boldsymbol{\nu}(S)\, \boldsymbol{\beta}(T) .$$

The proof of part (2) proceeds analogously.

**3.1.8. Proposition.** *For each nuclear mapping $T \in \mathcal{N}(E, F)$ the dual mapping $T'$ is also nuclear and $\boldsymbol{\nu}(T') \leqq \boldsymbol{\nu}(T)$.*

*Proof.* If $\delta$ is an arbitrary positive number, then the nuclear mapping $T$ can be expressed as

$$T x = \sum_N \langle x, a_n \rangle y_n \quad \text{for } x \in E$$

with

$$\sum_N p_{U^\circ}(a_n)\, p_V(y_n) \leqq \boldsymbol{\nu}(T) + \delta .$$

Then the dual mapping $T'$ has the form

$$T' b = \sum_N \langle y_n, b \rangle a_n \quad \text{for } b \in F' ,$$

and

$$\boldsymbol{\nu}(T') \leqq \sum_N p_{U^\circ}(a_n)\, p_V(y_n) \leqq \boldsymbol{\nu}(T) + \delta .$$

Our assertion is thus proved.

**3.1.9. Problem.** *Must each continuous mapping $T \in \mathcal{L}(E, F)$ whose dual mapping $T'$ is nuclear also be nuclear?*

Grothendieck has shown that the answer to this question is positive for a very wide class of normed spaces.

**3.1.10.** We now give some examples of nuclear mappings. For this purpose recall that each continuous linear mapping $T$ from $\boldsymbol{l}_I^1$ into $\boldsymbol{l}_J^1$

can be represented with uniquely determined matrix $[\tau_{ij}, I \times J]$ as

$$T[\xi_i, I] = \left[\sum_I \xi_i \tau_{ij}, J\right] \quad \text{for} \quad [\xi_i, I] \in \boldsymbol{l}_I^1 .$$

**Proposition.** *A mapping $T \in \mathscr{L}(\boldsymbol{l}_I^1, \boldsymbol{l}_J^1)$ is nuclear if and only if*

$$\boldsymbol{\varrho}(T) = \sum_J \sup\{|\tau_{ij}| : i \in I\} < +\infty .$$

*We then have $\boldsymbol{\nu}(T) = \boldsymbol{\varrho}(T)$.*

*Proof.* Since the mapping $T$ is nuclear, for each positive number $\delta$ there are families $a_n = [\alpha_i^{[n]}, I] \in \boldsymbol{m}_I$ and $y_n = [\eta_j^{[n]}, J] \in \boldsymbol{l}_J^1$ with

$$\sum_N \boldsymbol{\lambda}_\infty(a_n)\, \boldsymbol{\lambda}_1(y_n) \leqq \boldsymbol{\nu}(T) + \delta ,$$

so that for all families $x = [\xi_i, I] \in \boldsymbol{l}_I^1$, $T$ has the representation

$$T x = \sum_N \langle x, a_n \rangle\, y_n$$

or

$$T[\xi_i, I] = \left[\sum_N \left(\sum_I \alpha_i^{[n]} \xi_i\right) \eta_j^{[n]}, J\right] = \left[\sum_I \xi_i \left(\sum_N \alpha_i^{[n]} \eta_j^{[n]}\right), J\right].$$

Consequently, we have

$$\tau_{ij} = \sum_N \alpha_i^{[n]} \eta_j^{[n]} ,$$

and

$$\sup\{|\tau_{ij}| : i \in I\} \leqq \sum_N \boldsymbol{\lambda}_\infty(a_n)\, |\eta_j^{[n]}| .$$

Hence, we get the inequality

$$\boldsymbol{\varrho}(T) \leqq \sum_J \sum_N \boldsymbol{\lambda}_\infty(a_n)\, |\eta_j^{[n]}| = \sum_N \boldsymbol{\lambda}_\infty(a_n)\, \boldsymbol{\lambda}_1(y_n) \leqq \boldsymbol{\nu}(T) + \delta .$$

Therefore, for the nuclear mapping $T$, we have proved the relation $\boldsymbol{\varrho}(T) \leqq \boldsymbol{\nu}(T)$.

We now consider a continuous linear mapping $T$ from $\boldsymbol{l}_I^1$ into $\boldsymbol{l}_J^1$ with

$$\boldsymbol{\varrho}(T) = \sum_J \sup\{|\tau_{ij}| : i \in I\} < +\infty .$$

If we set

$$\varrho_j = \sup\{|\tau_{ij}| : i \in I\} ,$$

then the set $K = \{k \in J : \varrho_k > 0\}$ is at most countably infinite since

$$\sum_J \varrho_j < +\infty .$$

We define on $\boldsymbol{l}_I^1$ continuous linear forms $a_k$ by the equation

$$\langle [\xi_i, I], a_k \rangle = \sum_I \xi_i \tau_{ik}$$

and set

$$y_k = [\delta_j^{[k]}, J] .$$

Then the mapping $T$ has the form

$$T[\xi_i, I] = \left[\sum_I \xi_i \tau_{ij}, J\right] = \sum_K \left(\sum_I \xi_i \tau_{ik}\right) [\delta_j^{[k]}, J]$$

or

$$T x = \sum_K \langle y, a_k \rangle y_k$$

for each family $[\xi_i, I] \in \boldsymbol{l}_I^1$.

Since the relation

$$\sum_K \boldsymbol{\lambda}_\infty(a_k)\, \boldsymbol{\lambda}_1(y_k) = \sum_K \varrho_k = \boldsymbol{\varrho}(T) < +\infty$$

also holds, $T$ is nuclear and the inequality $\boldsymbol{\nu}(T) \leqq \boldsymbol{\varrho}(T)$ is valid.

## 3.2. Quasinuclear Mappings in Normed Spaces

**3.2.1.** If $F$ is a linear subspace of the normed space $G$, each continuous linear mapping from $E$ into $F$ can also be considered as a mapping from $E$ into $G$. It can thus happen that the mapping $T$ does not belong to $\mathcal{N}(E, F)$, although it is nuclear as a mapping from $E$ into $G$.

**3.2.2.** We now show that this pathological situation cannot occur if $F$ is dense in $G$.

**Lemma.** *If $F$ is a dense linear subspace of the normed space $G$ then for each element $z \in G$ there is a sequence of elements $y_n \in F$ with*

$$z = \sum_N y_n ,$$

*such that for an arbitrary positive number $\delta$ we have the inequality*

$$\sum_N p_W(y_n) \leqq (1 + \delta)\, p_W(z) .$$

*Proof.* By hypothesis there exist in $F$ elements $\hat{y}_n$ with

$$p_W(z - \hat{y}_n) \leqq (1/2^{n+1})\, \delta\, p_W(z) .$$

If we set $y_1 = \hat{y}_1$ and $y_n = \hat{y}_n - \hat{y}_{n-1}$ for $n > 1$ we then have

$$z = \lim_m \hat{y}_m = \sum_N y_n .$$

Moreover,

$$p_W(y_1) \leqq (1 + (1/4)\, \delta)\, p_W(z)$$

and

$$p_W(y_n) = (1/2^{n+1} + 1/2^n)\, \delta\, p_W(z) \quad \text{for} \quad n > 1 .$$

Consequently,

$$\sum_N p_W(y_n) \leqq (1 + (1/4)\,\delta + \delta \sum_{n=2}^{\infty} 1/2^{n+1} + \delta \sum_{n=2}^{\infty} 1/2^n)\, p_w(z)$$
$$\leqq (1 + \delta)\, p_W(z)\,.$$

It is easy to see that each nuclear mapping from $E$ into $F$ is also a nuclear mapping from $E$ into $G$. Thus $T$ has $\boldsymbol{\nu}$-norm in $\mathcal{N}(E, F)$ as well as in $\mathcal{N}(E, G)$, which we shall differentiate by writing $\boldsymbol{\nu}^F(T)$ resp. $\boldsymbol{\nu}^G(T)$.

These preliminary remarks lead to the

**Proposition.** *If $F$ is a dense linear subspace of the normed space $G$, then for each continuous mapping $T$ of a normed space $E$ into $F$, $T \in \mathcal{N}(E, G)$ always implies $T \in \mathcal{N}(E, F)$ and we have $\boldsymbol{\nu}^F(T) = \boldsymbol{\nu}^G(T)$.*

*Proof.* Since $T$ is a nuclear mapping from $E$ into $G$, for each positive number $\delta$ there are $a_n \in E'$ and $z_n \in G$ with

$$\sum_N p_{U^\circ}(a_n)\, p_W(z_n) \leqq \boldsymbol{\nu}^G(T) + \delta$$

such that $T$ has the representation

$$T x = \sum_{n=1}^{\infty} \langle x, a_n \rangle\, z_n \qquad \text{for } x \in E\,.$$

If we now determine elements $y_{m\,n} \in F$ with

$$z_n = \sum_{m=1}^{\infty} y_{m\,n} \quad \text{and} \quad \sum_{m=1}^{\infty} p_W(y_{m\,n}) \leqq (1 + \delta)\, p_W(z_n)\,,$$

for the linear forms $a_{m\,n} = a_n$ for $m = 1, 2, \ldots$, we have the relation

$$T x = \sum_{m=1}^{\infty} \sum_{n=1}^{\infty} \langle x, a_{m\,n} \rangle\, y_{m\,n} \qquad \text{for } x \in E\,.$$

In addition

$$\boldsymbol{\nu}^F(T) \leqq \sum_{m=1}^{\infty} \sum_{n=1}^{\infty} p_{U^\circ}(a_{m\,n})\, p_W(y_{m\,n}) \leqq (1 + \delta)\,(\boldsymbol{\nu}^G(T) + \delta)\,.$$

Consequently, $T$ is also nuclear as a mapping from $E$ into $F$ and $\boldsymbol{\nu}^F(T) \leqq \boldsymbol{\nu}^G(T)$. Thus our assertion is completely proved since the inequality $\boldsymbol{\nu}^F(T) \geqq \boldsymbol{\nu}^G(T)$ results immediately from the definition of the $\boldsymbol{\nu}$-norm.

**3.2.3.** If $E$ and $F$ are two arbitrary normed spaces with closed unit balls $U$ and $V$, then we designate a continuous linear mapping $T$ from $E$ into $F$ as **quasinuclear** if there is a sequence of linear forms $a_n \in E'$ with

$$\sum_N p_{U^\circ}(a_n) < +\infty$$

such that

$$p_V(T\,x) \leqq \sum_N |\langle x, a_n\rangle| \quad \text{for} \quad x \in E\,.$$

For each quasinuclear mapping $T$ we set

$$\boldsymbol{\pi}_0(T) = \inf\left\{\sum_N p_{U^0}(a_n)\right\},$$

where the infimum is taken over all sequences of linear forms $a_n$ which have the stated property.

**3.2.4.** It is evident that quasinuclear mappings do not have the undesirable properties mentioned in 3.2.1.

**3.2.5.** In the sequel we shall investigate the relation between nuclear and quasinuclear mappings. As our first result we have the

**Proposition.** *Each nuclear mapping $T$ is quasinuclear and* $\boldsymbol{\pi}_0(T) \leqq \boldsymbol{\nu}(T)$.

*Proof.* We consider an arbitrary nuclear mapping $T \in \mathcal{N}(E, F)$. For each positive number $\delta$ there are linear forms $a_n \in E'$ and elements $y_n \in F$ with

$$\sum_N p_{U^0}(a_n)\, p_V(y_n) \leqq \boldsymbol{\nu}(T) + \delta$$

such that $T$ has the form

$$T\,x = \sum_N \langle x, a_n\rangle\, y_n \quad \text{for} \quad x \in E\,.$$

If we set $b_n = p_V(y_n)\, a_n$, we get

$$p_V(T\,x) \leqq \sum_N |\langle x, a_n\rangle|\, p_V(y_n) = \sum_N |\langle x, b_n\rangle| \quad \text{for} \quad x \in E\,.$$

Because of the inequality

$$\sum_N p_{U^0}(b_n) \leqq \sum_N p_{U^0}(a_n)\, p_V(y_n) \leqq \boldsymbol{\nu}(T) + \delta\,,$$

$T$ has to be a quasinuclear mapping with $\boldsymbol{\pi}_0(T) \leqq \boldsymbol{\nu}(T)$.

**3.2.6.** A few preliminaries are needed for the proof of the theorem which follows.

**Lemma 1.** *Each normed space $F$ can be considered as a linear subspace of a Banach space* $\boldsymbol{m}_I$.

*Proof.* We consider an index set $I$ such that a mapping $i \to b_i$ exists from $I$ onto $V^0$. For each element $y \in F$ we determine the family $[\langle y, b_i\rangle, I] \in \boldsymbol{m}_I$, such that

$$p_V(y) = \sup\{|\langle y, b_i\rangle| : i \in I\} = \boldsymbol{\lambda}_\infty[\langle y, b_i\rangle, I]\,.$$

Therefore, $F$ can be regarded as a linear subspace of $\boldsymbol{m}_I$.

**Lemma 2.** *If $G_0$ is a linear subspace of the normed space $G$, then each continuous linear mapping $S_0$ of $G_0$ into a Banach space $\boldsymbol{m}_I$ can be extended to a continuous linear mapping $S$ from $G$ into $\boldsymbol{m}_I$ with $\boldsymbol{\beta}(S) = \boldsymbol{\beta}(S_0)$.*

*Proof.* The mapping $S_0$ can be written by means of a family of continuous linear forms $c_0^{[i]} \in G_0'$ as

$$S_0 z = [\langle z, c_0^{[i]} \rangle, I] \quad \text{for} \quad z \in G_0 .$$

Here we have $p_{W^0}(c_0^{[i]}) \leqq \boldsymbol{\beta}(S_0)$.

By the Hahn-Banach-Theorem 0.4.4 the linear forms $c_0^{[i]}$ can be extended to linear forms $c^{[i]}$ on all of $G$ without altering their norm. We now obtain the desired mapping $S$ from the equation

$$S z = [\langle z, c^{[i]} \rangle, I] \quad \text{for} \quad z \in G .$$

These preliminaries lead to the following

**Theorem.** *Each quasinuclear mapping $T$ from a normed space $E$ into a normed space $F$ is also nuclear if it is regarded as a mapping from $E$ into a Banach space $\boldsymbol{m}_I$ containing $F$. When this occurs $\boldsymbol{\pi}_0(T) = \boldsymbol{\nu}(T)$.*

*Proof.* Since the mapping $T$ is quasinuclear, there is for each positive number $\delta$ a sequence of linear forms $a_n \in E'$ with

$$\sum_N p_{U^0}(a_n) \leqq \boldsymbol{\pi}_0(T) + \delta ,$$

such that

$$p_V(T x) \leqq \sum_N |\langle x, a_n \rangle| \quad \text{for} \quad x \in E .$$

If $G_0$ as a subspace of $\boldsymbol{l}_N^1$ is formed of all families $[\langle x, a_n \rangle, N]$ with $x \in E$, then by the equation

$$S_0 [\langle x, a_n \rangle, N] = T x$$

we obtain a continuous linear mapping $S_0$ from $G_0$ into $F$ for which $\boldsymbol{\beta}(S_0) \leqq 1$ because

$$p_V(S_0 [\langle x, a_n \rangle, N]) = p_V(T x) \leqq \sum_N |\langle x, a_n \rangle| = \boldsymbol{\lambda}_1[\langle x, a_n \rangle, N] .$$

By Lemma 2 this mapping can be extended to a continuous linear mapping $S$ from $\boldsymbol{l}_N^1$ into $\boldsymbol{m}_I$ with $\boldsymbol{\beta}(S) \leqq 1$.

If we set

$$y_m = S[\delta_n^{[m]}, N] ,$$

then $\boldsymbol{\lambda}_\infty[y_m] \leqq 1$ and the mapping $S$ has the representation

$$S[\xi_n, N] = \sum_N \xi_n y_n \quad \text{for} \quad [\xi_n, N] \in \boldsymbol{l}_N^1 .$$

Since $T x = S[\langle x, a_n\rangle, N]$, the mapping $T$ has the form

$$T x = \sum_N \langle x, a_n\rangle y_n \quad \text{for} \quad x \in E .$$

Because of the inequality

$$\sum_N p_{U^\circ}(a_n)\, \boldsymbol{\lambda}_\infty[y_n] \leqq \boldsymbol{\pi}_0(T) + \delta ,$$

$T$ must be nuclear as a mapping from $E$ into $\boldsymbol{m}_I$. Moreover, since $\boldsymbol{\nu}(T) \leqq \boldsymbol{\pi}_0(T) + \delta$, we have $\boldsymbol{\nu}(T) \leqq \boldsymbol{\pi}_0(T)$ by taking the limit as $\delta \to 0$. By Proposition 3.2.5, the identity $\boldsymbol{\nu}(T) = \boldsymbol{\pi}_0(T)$ is also true.

**3.2.7.** A combination of the results of 3.2.5 and 3.2.6 gives the

**Proposition.** *A continuous linear mapping $T$ from a normed space $E$ into a normed space $F$ is quasinuclear if and only if there is a normed space $G$ containing $F$ such that $T$ is nuclear as a mapping from $E$ into $G$.*

**3.2.8.** If the collection of all quasinuclear mappings of $E$ into $F$ is designated as $\mathcal{P}_0(E, F)$, then we have the

**Proposition.** *$\mathcal{P}_0(E, F)$ is a linear space with the norm $\boldsymbol{\pi}_0(T)$.*

*Proof.* Our assertion is derived immediately from the fact that $\mathcal{P}_0(E, F)$ can be regarded as a linear subspace of $\mathcal{N}(E, \boldsymbol{m}_I)$ by Theorem 3.2.6.

**3.2.9.** By an argument like that in 3.2.8 we obtain the

**Proposition.** *For a normed space $E$ and a Banach space $F$, $\mathcal{P}_0(E, F)$ is a Banach space.*

**3.2.10. Proposition.** *Each quasinuclear mapping is precompact.*

**Proof.** Since each mapping $T \in \mathcal{P}_0(E, F)$ is nuclear and thus precompact as a mapping from $E$ into $\boldsymbol{m}_I$ it must also be precompact as a mapping from $E$ into $F$.

**3.2.11.** As an immediate consequence of 3.2.10 and 0.10.6 we obtain the

**Proposition.** *Each quasinuclear mapping has a separable range.*

**3.2.12.** Let $E$, $F$ and $G$ be three normed spaces. We then have the

**Proposition.**

(1) *If $T \in \mathcal{L}(E, F)$ and $S \in \mathcal{P}_0(F, G)$, then $S\, T \in \mathcal{P}_0(E, G)$ and* $\boldsymbol{\pi}_0(S\, T) \leqq \boldsymbol{\pi}_0(S)\, \boldsymbol{\beta}(T)$ .

(2) *If $T \in \mathcal{P}_0(E, F)$ and $S \in \mathcal{L}(F, G)$, then $S\, T \in \mathcal{P}_0(E, G)$ and* $\boldsymbol{\pi}_0(S\, T) \leqq \boldsymbol{\beta}(S)\, \boldsymbol{\pi}_0(T)$.

**3.2.13.** Finally, we investigate the relation between quasinuclear and absolutely summing mappings.

**Proposition.** *Every quasinuclear mapping $T$ is absolutely summing and* $\pi(T) \leqq \pi_0(T)$.

*Proof.* We consider an arbitrary quasinuclear mapping $T \in \mathcal{P}_0(E,F)$. Then for each positive number $\delta$ there is a sequence of linear forms $a_n \in E'$ with

$$\sum_N p_{U^0}(a_n) \leqq \pi_0(T) + \delta ,$$

such that

$$p_V(T\,x) \leqq \sum_N |\langle x, a_n\rangle| \quad \text{for} \quad x \in E .$$

With

$$\mu_n = p_{U^0}(a_n) \quad \text{and} \quad b_n = \begin{cases} \mu_n^{-1}\, a_n & \text{for } \mu_n > 0 \\ 0 & \text{for } \mu_n = 0 \end{cases}$$

we define a positive Radon measure $\mu$ on $U^0$ by

$$\langle \varphi, \mu\rangle = \int_{U^0} \varphi(a)\, d\mu = \sum_N \mu_n\, \varphi(b_n) \quad \text{for} \quad \varphi \in \mathscr{C}(U^0)$$

such that

$$p_V(T\,x) \leqq \int_{U^0} |\langle x, a\rangle|\, d\mu \quad \text{for all} \quad x \in E .$$

Consequently, $T$ is absolutely summing by Theorem 2.3.3.

Finally from the inequality

$$\pi(T) \leqq \mu(U^0) = \sum_N \mu_n \leqq \pi_0(T) + \delta$$

we obtain the assertion $\pi(T) \leqq \pi_0(T)$ by taking the limit as $\delta \to 0$.

**3.2.14.** It can be shown that for each quasinuclear mapping $T$, the identity

$$\pi_0(T) = \pi(T)$$

is valid. Moreover, for a wide class of normed spaces, $\mathcal{P}_0(E, F)$ consists precisely of those absolutely summing mappings which can be approximated by finite mappings with respect to the $\pi$-norm. Theorem 2.5.5. can thus be strengthened to the statement that all Hilbert-Schmidt-mappings are quasinuclear. The proof of these results is found in Pietsch [8].

## 3.3. Products of Quasinuclear and Absolutely Summing Mappings in Normed Spaces

**3.3.1.** We begin the investigations of this section with the following

**Lemma.** *Each quasinuclear mapping $T$ of a normed space $E$ into a Banach space $F$ can be represented as the product of two continuous linear mappings in the following way:*

$$E \xrightarrow{T_1} \boldsymbol{m}_N \xrightarrow{T_2} F$$

*where for each positive number $\delta$ we have the inequalities*

$$\boldsymbol{\beta}(T_1) \leqq 1 \quad \textit{and} \quad \boldsymbol{\beta}(T_2) \leqq \boldsymbol{\pi}_0(T) + \delta .$$

*Proof.* We determine a sequence of linear forms $a_n \in E'$ with

$$\sum_N p_{U^0}(a_n) \leqq \boldsymbol{\pi}_0(T) + \delta ,$$

so that

$$p_V(T\,x) \leqq \sum_N |\langle x, a_n\rangle| \quad \text{for all} \quad x \in E .$$

If $N_0$ is the set of those natural numbers $n$ with $a_n \neq$ o, we set

$$\mu_n = p_{U^0}(a_n)^{1/2} \quad \text{and} \quad b_n = \mu_n^{-2}\, a_n \quad \text{for} \quad n \in N_0 .$$

Now we define the mappings $T_1 \in \mathscr{L}(E, \boldsymbol{m}_N)$ and $S_1 \in \mathscr{L}(\boldsymbol{m}_N, \boldsymbol{l}^2_{N_0})$ by the equations

$$T_1\, x = [\langle x, b_n\rangle, N] \quad \text{for} \quad x \in E$$

and

$$S_1[\xi_n, N] = [\mu_n\, \xi_n, N_0] \quad \text{for} \quad [\xi_n, N] \in \boldsymbol{m}_N .$$

Since

$$\boldsymbol{\lambda}_\infty(T_1\, x) = \sup\{|\langle x, b_n\rangle| : n \in N\} \leqq p_U(x)$$

and

$$\boldsymbol{\lambda}_2(S_1[\xi_n, N_0]) = \left\{\sum_{N_0} \mu_n^2\, |\xi_n|^2\right\}^{1/2} \leqq \left\{\sum_{N_0} \mu_n^2\right\}^{1/2} \boldsymbol{\lambda}_\infty[\xi_n, N]$$

we have

$$\boldsymbol{\beta}(T_1) \leqq 1 \quad \text{and} \quad \boldsymbol{\beta}(S_1) \leqq \{\boldsymbol{\pi}_0(T) + \delta\}^{1/2} .$$

Now we define on the linear subspace

$$H = \{[\mu_n \langle x, b_n\rangle, N_0] : x \in E\}$$

of the Hilbert space $\boldsymbol{l}^2_{N_0}$, a continuous linear mapping $S_2$ by

$$S_2\, [\mu_n \langle x, b_n\rangle, N_0] = T\, x .$$

For this mapping $S_2$ we have the inequality

$$\boldsymbol{\beta}(S_2) \leqq \{\boldsymbol{\pi}_0(T) + \delta)\}^{1/2}$$

because

$$p_V(S_2\, [\mu_n \langle x, b_n\rangle, N_0]) = p_V(T\, x) \leqq \sum_{N_0} \mu_n^2\, |\langle x, b_n\rangle|$$
$$\leqq \left\{\sum_{N_0} \mu_n^2\right\}^{1/2} \boldsymbol{\lambda}_2[\langle \mu_n\, x_n, b_n\rangle, N_0] .$$

By 0.10.7 there is a projection $P$ on the Hilbert space $\boldsymbol{l}^2_{N_0}$ with $\mathfrak{R}(P) = \overline{H}$ so that we can set $T_2 = \overline{S_2}\, P\, S_1$ where $\overline{S_2}$ is the uniquely defined continuous extension of $S_2$ to the closed hull of $H$. The norm of the mapping $T_2$ satisfies the inequality

$$\boldsymbol{\beta}(T_2) = \boldsymbol{\beta}(\overline{S_2})\, \boldsymbol{\beta}(P)\, \boldsymbol{\beta}(S_1) \leqq \boldsymbol{\pi}_0(T) + \delta .$$

Thus our conclusion is proved since the mapping $T$ has the form $T = T_2 T_1$. This is because

$$T x = S_2[\mu_n\langle x, b_n\rangle, N_0] = S_2 P S_1[\langle x, b_n\rangle, N] = T_2 T_1 x .$$

**3.3.2.** For $E, F$ and $G$, three normed spaces we have the

**Theorem.** *The product $S T$ of two quasinuclear mappings $T \in \mathcal{P}_0(E, F)$ and $S \in \mathcal{P}_0(F, G)$ is nuclear and*

$$\boldsymbol{\nu}(S T) \leqq \boldsymbol{\pi}_0(S) \boldsymbol{\pi}_0(T) .$$

*Proof.* We consider $S$ as a mapping from $F$ into $\tilde{G}$. Then, by the previous lemma, for each positive number $\delta$ there are two mappings

$$S_1 \in \mathcal{L}(F, \boldsymbol{m}_N) \quad \text{and} \quad S_2 \in \mathcal{L}(\boldsymbol{m}_N, \tilde{G})$$

with $\boldsymbol{\beta}(S_1) \leqq 1$ and $\boldsymbol{\beta}(S_2) \leqq \boldsymbol{\pi}_0(S) + \delta$.

Now from Theorem 3.2.6 and Proposition 3.2.12 it follows that $S_1 T$ is a nuclear mapping from $E$ into $\boldsymbol{m}_N$ with $\boldsymbol{\nu}(S_1 T) \leqq \boldsymbol{\pi}_0(T)$. Consequently, the product $S T = S_2 S_1 T$ must be a nuclear mapping from $E$ into $\tilde{G}$. Here we have

$$\boldsymbol{\nu}^{\tilde{G}}(S T) \leqq \boldsymbol{\beta}(S_2) \boldsymbol{\nu}(S_1 T) \leqq (\boldsymbol{\pi}_0(S) + \delta) \boldsymbol{\pi}_0(T) ,$$

and by taking the limit as $\delta \to 0$ we obtain

$$\boldsymbol{\nu}^{\tilde{G}}(S T) \leqq \boldsymbol{\pi}_0(S) \boldsymbol{\pi}_0(T) .$$

Our assertion is hereby proved, since by Proposition 3.3.2, $S T$ is also nuclear as a mapping from $E$ into $G$ and $\boldsymbol{\nu}^{\tilde{G}}(S T) = \boldsymbol{\nu}^{G}(S T)$.

**3.3.3.** In the sequel $M$ is an arbitrary compact Hausdorff space on which a positive Radon measure $\mu$ with $\mu(M) = 1$ is given. For the canonical mapping $K$ from $\mathscr{C}(M)$ into $\mathfrak{L}_\mu^2(M)$ defined by

$$K[\varphi] = \hat{\varphi}$$

we have $\boldsymbol{\beta}(K) = 1$.

**Proposition 1.** *For each continuous linear mapping $T$ from a Hilbert space $E$ into the Banach space $\mathscr{C}(M)$, $K T$ is a Hilbert-Schmidt-mapping from $E$ into $\mathfrak{L}_\mu^2(M)$ and $\boldsymbol{\sigma}(K T) \leqq \boldsymbol{\beta}(T)$.*

*Proof.* We consider a complete orthonormal system $[e_i, I]$ in $E$ and set $\varphi_i = T e_i$. If $\delta_x$ is the Dirac measure defined for the point $x \in M$ by the equation

$$\langle \varphi, \delta_x\rangle = \varphi(x) \quad \text{for} \quad \varphi \in \mathscr{C}(M) ,$$

then by the Bessel inequality

$$\sum_I |\varphi_i(x)|^2 = \sum_I |\langle T e_i, \delta_x\rangle|^2 = \sum_I |\langle e_i, T' \delta_x\rangle|^2 \leqq p_{U^0}(T' \delta_x)^2 \leqq \boldsymbol{\beta}(T)^2$$

and by integration over $M$ we obtain

$$\boldsymbol{\sigma}(K\,T)^2 = \sum_I \int_M |\varphi_i(x)|^2\,d\mu \leqq \boldsymbol{\beta}(T)^2\,,$$

which proves the proposition.

**Proposition 2.** *For each Hilbert-Schmidt-mapping $T$ from $\mathfrak{L}^2_\mu(M)$ into a Hilbert space $F$, $T\,K$ is a nuclear mapping from the Banach space $\mathscr{C}(M)$ into $F$ and $\boldsymbol{\nu}(T\,K) \leqq \boldsymbol{\sigma}(T)$.*

*Proof.*

(1) We first treat the case of a finite mapping $T$ which can be described in the form

$$T[\hat{f}] = \sum_{r=1}^{m} (\hat{f}, \hat{f}_r)\, y_r \quad \text{for } \hat{f} \in \mathfrak{L}^2_\mu(M)\,,$$

where $f_1, \ldots, f_m$ are step functions and $y_1, \ldots, y_m \in F$. Then there are finitely many disjoint $\mu$-measurable subsets $M_1, \ldots, M_n$ of $M$ with the property that the step functions $f_1, f_2, \ldots, f_m$ are linear combinations of the characteristic functions belonging to them:

$$f_r = \sum_{s=1}^{n} \alpha_{rs}\, g_s \quad \text{for } r = 1, \ldots, m\,.$$

For the elements

$$z_s = \sum_{r=1}^{m} \bar{\alpha}_{rs}\, y_r \qquad s = 1, \ldots, n$$

we have the identity

$$T[\hat{f}] = \sum_{s=1}^{n} (\hat{f}, \hat{g}_s)\, z_s \quad \text{for } \hat{f} \in \mathfrak{L}^2_\mu(M)\,.$$

If the positive Radon measures $\mu_s$ are defined by

$$\langle \varphi, \mu_s \rangle = \int_{M_s} \varphi(x)\, d\mu = (\hat{\varphi}, \hat{g}_s) \quad \text{for } \varphi \in \mathscr{C}(M)\,,$$

then the mapping $T\,K$ can be written as

$$T\,K[\varphi] = \sum_{s=1}^{n} \langle \varphi, \mu_s \rangle\, z_s \quad \text{with } \varphi \in \mathscr{C}(M)\,.$$

Therefore

$$\boldsymbol{\nu}(T\,K) \leqq \sum_{s=1}^{n} \mu(M_s)\, p_V(z_s) \leqq \left\{\sum_{s=1}^{n} \mu(M_s)\right\}^{1/2} \left\{\sum_{s=1}^{n} \mu(M_s)\, p_V(z_s)^2\right\}^{1/2}.$$

Since the function classes $\hat{h}_s = \mu(M_s)^{-1/2}\,\hat{g}_s$ form an orthogonal system because $(\hat{g}_s, \hat{g}_s) = \mu(M_s)$, we have

$$\boldsymbol{\sigma}(T)^2 \geqq \sum_{s=1}^{n} p_V(T[\hat{h}_s])^2 = \sum_{s=1}^{n} \mu(M_s)\, p_V(z_s)^2\,.$$

Because

$$\sum_{s=1}^{n} \mu(M_s) \leqq \mu(M) = 1 ,$$

we have $\boldsymbol{\nu}(T\,K) \leqq \boldsymbol{\sigma}(T)$ and the proposition holds for this special case.

(2) If $T$ is an arbitrary Hilbert-Schmidt-mapping from $\mathfrak{L}^2_\mu(M)$ into $F$, then for a complete orthonormal system $[e_i, I]$ of $F$ we get

$$\boldsymbol{\sigma}(T)^2 = \sum_I \boldsymbol{\lambda}_2(T^*\, e_i)^2 < +\infty .$$

Consequently, for each natural number $n$ there is a set $\mathfrak{i}_n \in \mathfrak{F}(I)$ with

$$\sum_{I \setminus \mathfrak{i}_n} \boldsymbol{\lambda}_2(T^*\, e_i)^2 < 1/(2\, n^2) .$$

We determine step functions $\hat{f}_i^{[n]}$ with

$$\boldsymbol{\lambda}_2(T^*\, e_i - \hat{f}_i^{[n]})^2 < 1/(2|\mathfrak{i}_n| n^2)$$

where $|\mathfrak{i}_n|$ is the number of elements in $\mathfrak{i}_n$. Finally, if we define the mappings $T_n$ by the equation

$$T_n[\hat{f}] = \sum_{\mathfrak{i}_n} (\hat{f}, \hat{f}_i^{[n]})\, e_i \quad \text{for } \hat{f} \in \mathfrak{L}^2_\mu(M)$$

then, since

$$\boldsymbol{\sigma}\,(T - T_n)^2 = \sum_{\mathfrak{i}_n} \boldsymbol{\lambda}_2(T^*\, e_i - \hat{f}_i^{[n]})^2 + \sum_{I \setminus \mathfrak{i}_n} \boldsymbol{\lambda}_2(T\, e_i)^2 ,$$

we get

$$\boldsymbol{\sigma}(T - T_n) \leqq 1/n .$$

Therefore

$$\boldsymbol{\sigma} - \lim_n T_n = T .$$

By applying the special case (1) to the mapping $T_m - T_n$, we obtain the inequality

$$\boldsymbol{\nu}(T_m\, K - T_n\, K) \leqq \boldsymbol{\sigma}\,(T_m - T_n) \quad \text{for } m, n \in N .$$

Therefore, the mappings $T_n\, K$ form a $\boldsymbol{\nu}$-Cauchy-sequence. Since

$$p_V(T\, K[\varphi] - T_n\, K[\varphi]) \leqq \boldsymbol{\beta}\,(T - T_n)\, \boldsymbol{\lambda}_2\,(K[\varphi]) \leqq \boldsymbol{\sigma}(T - T_n)\, \|\varphi\| ,$$

we have

$$\lim_n T_n\, K[\varphi] = T\, K[\varphi] \quad \text{for } \varphi \in \mathscr{C}(M)$$

and so by Lemma 3.1.3 the mapping $T\,K$ is nuclear. Also

$$\boldsymbol{\nu}(T\, K) = \lim_n \boldsymbol{\nu}(T_n\, K) \leqq \lim_n \boldsymbol{\sigma}(T_n) = \boldsymbol{\sigma}(T) .$$

**3.3.4.** We now give a representation of absolutely summing mappings which is analogous to Lemma 3.3.1.

**Lemma.** *Each absolutely summing mapping $T$ from a normed space $E$ into a Banach space $F$ can be described as the product of three continuous linear mappings as follows:*

$$E \xrightarrow{T_1} \mathscr{C}(M) \xrightarrow{K} \mathfrak{L}^2_\mu(M) \xrightarrow{T_2} F .$$

*Here, $\mu$ is a positive Radon measure on a compact Hausdorff space $M$ for which $\mu(M) = 1$ and the mapping $K$ is defined by the equation $K[\varphi] = \hat{\varphi}$ for $\varphi \in \mathscr{C}(M)$. We also have*

$$\boldsymbol{\beta}(T_1) \leqq 1 \quad \textit{and} \quad \boldsymbol{\beta}(T_2) \leqq \boldsymbol{\pi}(T) .$$

*Proof.* Without loss of generality we can suppose that $\boldsymbol{\pi}(T) > 0$ since the case $T = 0$ is trivial. Then by Theorem 2.3.3 there is a positive Radon measure $\lambda$ on the set $M = U^0$ with $\lambda(M) = \boldsymbol{\pi}(T)$ such that

$$p_V(T\,x) \leqq \int_M |\langle x, a\rangle|\, d\lambda \quad \text{for all } x \in E .$$

Since the function $\varphi_x$ with $\varphi_x(a) = \langle x, a\rangle$ belongs to $\mathscr{C}(M)$ for each element $x \in E$, the equation

$$T_1\, x = \varphi_x \quad \text{for } x \in E$$

defines a continuous linear mapping $T_1$ from $E$ into $\mathscr{C}(M)$ for which we have $\boldsymbol{\beta}(T_1) = 1$ because

$$\|\varphi_x\| = \sup\{|\langle x, a\rangle| : a \in M\} = p_U(x) .$$

We normalize the positive Radon measure $\lambda$ by setting

$$\int_M \varphi(a)\, d\mu = \boldsymbol{\pi}(T)^{-1} \int_M \varphi(a)\, d\lambda \quad \text{for } \varphi \in \mathscr{C}(M) .$$

Then we get the estimate

$$p_V(T\,x) \leqq \boldsymbol{\pi}(T) \int_M |\langle x, a\rangle|\, d\mu \quad \text{for } x \in E .$$

We now define on the linear subspace

$$H = \{\hat{\varphi}_x : x \in E\}$$

of the Hilbert space $\mathfrak{L}^2_\mu(M)$, a continuous linear mapping $S_2$ by

$$S_2(\hat{\varphi}_x) = T\,x$$

for which we have the inequality $\boldsymbol{\beta}(S_2) \leqq \boldsymbol{\pi}(T)$ because

$$p_V[S_2(\hat{\varphi}_x)] \leqq \boldsymbol{\pi}(T) \int_M |\varphi_x(a)|\, d\mu \leqq \boldsymbol{\pi}(T) \left\{\int_M |\varphi_x(a)|^2\, d\mu\right\}^{1/2} \leqq \boldsymbol{\pi}(T)\, \boldsymbol{\lambda}_2(\hat{\varphi}_x) .$$

Since by 0.10.7 there is a projection $P$ on the Hilbert space $\mathfrak{L}^2_\mu(M)$ with $\mathfrak{R}(P) = \overline{H}$, we can set $T_2 = \overline{S}_2\, P$ where $\overline{S}_2$ is the uniquely defined continuous extension of $S_2$ to the closed hull of $H$. The norm of $T_2$ then satisfies the inequality $\boldsymbol{\beta}(T_2) \leqq \boldsymbol{\pi}(T)$.

The lemma is thus proved since because of

$$T\,x = S_2(\hat{\varphi}_x) = S_2\,P(\hat{\varphi}_x) = T_2\,K(\varphi_x) = T_2\,K\,T_1\,x\,,$$

the mapping $T$ has the form $T = T_2\,K\,T_1$.

**3.3.5.** For $E$, $F$, and $G$ three normed spaces, we now obtain the following fundamental theorem which is a generalization of the result of 3.3.2.

**Theorem.** *The product $S\,T$ of two absolutely summing mappings $T \in \mathfrak{P}(E, F)$ and $S \in \mathfrak{P}(F, G)$ is nuclear and*

$$\boldsymbol{\nu}(S\,T) \leqq \boldsymbol{\pi}(S)\,\boldsymbol{\pi}(T)\,.$$

*Proof.* By the preceding lemma the mappings $S$ and $T$ can be decomposed in the following way:

$$E \xrightarrow{T_1} \mathscr{C}(M_T) \xrightarrow{K_T} \mathfrak{L}^2_\mu(M_T) \xrightarrow{T_2} \tilde{F}\,.$$

$$F \xrightarrow{S_1} \mathscr{C}(M_S) \xrightarrow{K_S} \mathfrak{L}^2_\mu(M_S) \xrightarrow{S_2} \tilde{G}\,.$$

Then if $\tilde{S}_1$ is the continuous extension of the mapping $S_1$ to the complete hull of $F$, we obtain for $S\,T$ as a mapping from $E$ into $\tilde{G}$ the following product representation:

$$S\,T = S_2\,K_S\,\tilde{S}_1\,T_2\,K_T\,T_1\,.$$

Because of 3.3.3 (Proposition 1), $K_S(\tilde{S}_1\,T_2)$ is a Hilbert-Schmidt-mapping from $\mathfrak{L}^2_\mu(M_T)$ into $\mathfrak{L}^2_\lambda(M_S)$ and we have

$$\boldsymbol{\sigma}(K_S\,\tilde{S}_1\,T_2) \leqq \boldsymbol{\beta}(\tilde{S}_1\,T_2) \leqq \boldsymbol{\pi}(T)\,.$$

Now it follows from 3.3.3 (Proposition 2) that the mapping $(K_S\tilde{S}_1\,T_2)\,K_T$ is nuclear. Moreover,

$$\boldsymbol{\nu}(K_S\,\tilde{S}_1\,T_2\,K_T) \leqq \boldsymbol{\sigma}(K_S\,\tilde{S}_1\,T_2) \leqq \boldsymbol{\pi}(T)\,.$$

Consequently, $S\,T$ is nuclear as a mapping from $E$ into $\tilde{G}$ and

$$\boldsymbol{\nu}^{\tilde{G}}(S\,T) \leqq \boldsymbol{\beta}(S_2)\,\boldsymbol{\nu}(K\,\tilde{S}_1\,T_2\,K_T)\,\boldsymbol{\beta}(T_1) \leqq \boldsymbol{\pi}(S)\,\boldsymbol{\pi}(T)\,.$$

Our assertion is therefore proved since by Proposition 3.2.2, $S\,T$ is also a nuclear mapping from $E$ into $G$ and $\boldsymbol{\nu}^G(S\,T) = \boldsymbol{\nu}^{\tilde{G}}(S\,T)$.

**3.3.6.** For $E$, $F$ and $G$, three Hilbert spaces, the previous theorem can be proved more easily.

**Theorem.** *The product $S\,T$ of two Hilbert-Schmidt-mappings $T \in \mathfrak{L}(E, F)$ and $S \in \mathfrak{L}(F, G)$ is nuclear and*

$$\boldsymbol{\nu}(S\,T) \leqq \boldsymbol{\sigma}(S)\,\boldsymbol{\sigma}(T)\,.$$

*Proof.* If $[f_i, I]$ is a complete orthonormal system in $F$, then from the inequality

$$\sum_I p_W(S\, f_i)^2 < +\infty$$

it follows that the set

$$I_0 = \{i \in I : S\, f_i \neq 0\}\,,$$

is at most countable infinite.

From the identity

$$S\, y = \sum_{I_0} (y, f_i)\, S\, f_i \quad \text{for} \quad y \in F$$

we obtain the representation

$$S\, T\, x = \sum_{I_0} (x, T^*\, f_i)\, S\, f_i \quad \text{with} \quad x \in E\,.$$

If we now set $z_i = S\, f_i$ and if the continuous linear forms $a_i \in E'$ are defined by the formula

$$\langle x, a_i \rangle = (x, T^*\, f_i) \quad \text{for} \quad x \in E\,,$$

we then have

$$S\, T\, x = \sum_{I_0} \langle x, a_i \rangle\, z_i \quad \text{for} \quad x \in E\,.$$

Moreover, we also have

$$\boldsymbol{\nu}(S\, T) \leqq \sum_{I_0} p_{U^0}(a_i)\, p_W(z_i) = \sum_{I_0} p_U(T^*\, f_i)\, p_W(S\, f_i)$$

$$= \left\{\sum_{I_0} p_U(T^*\, f_i)^2\right\}^{1/2} \left\{\sum_{I_0} p_W(S\, f_i)^2\right\}^{1/2} = \boldsymbol{\sigma}(T)\, \boldsymbol{\sigma}(S)\,.$$

Consequently, $S\, T$ is a nuclear mapping with $\boldsymbol{\nu}(S\, T) \leqq \boldsymbol{\sigma}(S)\, \boldsymbol{\sigma}(T)$.

*Remark.* By applying Proposition 1 of 3.3.3, Lemma 3.3.4 and Theorem 3.3.6 we can easily show that in normed spaces the product of three absolutely summing mappings is always nuclear. This weakening of Theorem 3.3.5 is sufficient for all applications in the theory of nuclear locally convex spaces.

## 3.4. The Theorem of Dvoretzky and Rogers

**3.4.1.** As an application of our previous results we obtain the

**Dvoretzky Rogers Theorem.** *In each infinite dimensional normed space $E$ there are summable families $[x_n, N]$ which are not absolutely summable.*

*Proof*. We assume that in a normed space $E$ all summable families $[x_n, N]$ are also absolutely summable. Then the identity operator $K$ from $E$ into itself is absolutely summing. Then by 3.3.5 it follows $K = K^2$ is nuclear and thus, precompact. Therefore, the closed unit ball $U = K(U)$ of $E$ is precompact, and $E$ must be finite dimensional because of 0.8.3.

Chapter 4

# Nuclear Locally Convex Spaces

We begin this chapter with the defitinion of *nuclear locally convex spaces* which was introduced by A. Grothendieck [1] in 1951 in the setting of his theory of topological tensor products.

An important result is Theorem 3.2.4. It contains as a special case the following criterion of Grothendieck: *A metric locally convex space is nuclear if and only if every summable sequence in its is also absolutely summable.*

Of great significance is the question, under what assumptions the strong topological dual of a nuclear locally convex space is also nuclear. A necessary and sufficient condition for this to happen is given for the first time in Theorem 4.3.1. In order to better formulate results concerning this type of problem we shall in the sequel designate as *dual nuclear* all locally convex spaces whose strong topological dual is nuclear.

In section 4.4 various properties of nuclear spaces are collected.

## 4.1. Definition of Nuclear Locally Concvex Spaces

**4.1.1.** To initiate the theory of nuclear locally convex spaces we state the

**Lemma.** *For a system $\mathfrak{M}(E)$ of closed and absolutely convex bounded subsets of a locally convex space $E$ the following properties are equivalent:*

($N$) *For each set $A \in \mathfrak{M}(E)$ there is a set $B \in \mathfrak{M}(E)$ with $A < B$, such that the canonical mapping of $E(A)$ into $E(B)$ is nuclear, resp. quasinuclear, resp. absolutely summing.*

($N'$) *For each set $A \in \mathfrak{M}(E)$ there is a set $B \in \mathfrak{M}(E)$ with $A < B$ such that the canonical mapping from $E'(B^0)$ onto $E'(A^0)$ is nuclear, resp. quasinuclear, resp. absolutely summing.*

*Proof.* At the outset we establish that the properties collected under ($N$) are equivalent. In the first place, if we known that the canonical mapping from $E(A)$ into $E(B)$ is nuclear, then because of Propositions 3.2.5 and 3.2.13 it must be quasinuclear and absolutely summing as well. On the other hand, the canonical mapping from $E(A)$ into $E(C)$ is nuclear by 3.3.2 or 3.3.5 if for the set $A \in \mathfrak{M}(E)$ we determine the sets $B, C \in \mathfrak{M}(E)$ with $A < B < C$ such that the canonical mappings $E(A, B)$ and $E(B, C)$ are quasinuclear or even merely absolutely summing.

The equivalence of the conditions brought together in $(N')$ is obtained in a completely analogous way.

We shall now assume that for each set $A$ in $\mathfrak{M}(E)$ there is a set $B$ with $A < B$ such that the canonical mapping from $E(A)$ into $E(B)$ is nuclear. If the normed spaces $E'(A^0)$ and $E'(B^0)$ are now regarded as linear subspaces of $[E(A)]_b'$ and $[E(B)]_b'$ according to 0.11.4, we then know that the canonical mapping $E'(B^0, A^0)$ is the restriction of the mapping dual to $E(A, B)$ which is nuclear because of Proposition 3.1.8. Consequently, the canonical mapping from $E'(B^0)$ onto $E'(A^0)$ must at least be quasinuclear.

In a completely analogous way we prove that property $(N)$ follows from property $(N')$.

**4.1.2.** A locally convex space $E$ is called **nuclear** if it contains a fundamental system $\mathfrak{U}_{\mathfrak{F}}(E)$ of zero neighborhoods which has the following equivalent properties:

$(N')$ *For each zero neighborhoodd* $U \in \mathfrak{U}_{\mathfrak{F}}(E)$ *there exists a zero neighborhood* $V \in \mathfrak{U}_{\mathfrak{F}}(E)$ *with* $V < U$ *such that the canonical mapping from* $E(V)$ *onto* $E(U)$ *is nuclear, resp. quasinuclear, resp. absolutely summing.*

$(N)$ *For each zero neighborhood* $U \in \mathfrak{U}_{\mathfrak{F}}(E)$ *there exists a zero neighborhood* $V \in \mathfrak{U}_{\mathfrak{F}}(E)$ *with* $V < U$ *such that the canonical mapping from* $E'(U^0)$ *into* $E'(V^0)$ *is nuclear, resp. quasinuclear, resp. absolutely summing.*

The equivalence of these properties is obtained by applying the preceding lemma to the system of sets

$$\{U^0 : U \in \mathfrak{U}_{\mathfrak{F}}(E)\}$$

of the locally convex space $E_s'$.

**4.1.3.** We now show that the nuclearity of a locally convex space does not depend on the choice of a special fundamental system of zero neighborhoods. In fact, we have the

**Proposition.** *In a nuclear locally convex space each fundamental system of zero neighborhoods has the properties* $(N)$ *and* $(N')$.

*Proof.* If $E$ is a nuclear locally convex space, there exists by hypothesis a fundamental system $\mathfrak{U}_{\mathfrak{F}}^1(E)$ of zero neighborhoods with property $(N')$. For every other fundamental system $\mathfrak{U}_{\mathfrak{F}}^2(E)$ of zero neighborhoods there is then for every $U_2 \in \mathfrak{U}_{\mathfrak{F}}^2(E)$ a $U_1 \in \mathfrak{U}_{\mathfrak{F}}^1(E)$ with $U_1 < U_2$. We now determine a $V_1 \in \mathfrak{U}_{\mathfrak{F}}^1(E)$ with $V_1 < U_1$ so that the canonical mapping of $E(V_1)$ onto $E(U_1)$ is nuclear. Finally we choose in $\mathfrak{U}_{\mathfrak{F}}^2(E)$ a zero neighborhood $V_2$ with $V_2 < V_1$. Then the canonical mapping from $E(V_2)$ onto $E(U_2)$ is nuclear because

$$E(V_2, U_2) = E(U_1, U_2)\, E(V_1, U_1)\, E(V_2, V_1)\,,$$

and we have thus shown that the fundamental system $\mathfrak{U}_{\mathfrak{F}}^2(E)$ also has the equivalent properties $(N)$ and $(N')$.

**4.1.4.** We now give a characterization of nuclear locally convex spaces in which the normed spaces $E(U)$ or $E'(U^0)$ with $U \in \mathfrak{U}(E)$ are not used.

**Proposition.** *A locally convex space E is nuclear if and only if some, resp. each, fundamental system $\mathfrak{U}_{\mathfrak{F}}(E)$ of zero neighborhoods has the following property:*

*(Q) For each zero neighborhood $U \in \mathfrak{U}_{\mathfrak{F}}(E)$ there is a zero neighborhood $V \in \mathfrak{U}_{\mathfrak{F}}(E)$ and a sequence of continuous linear forms $a_n \in E'$ with*

$$\sum_N p_{V^0}(a_n) < +\infty ,$$

*such that the inequality*

$$p_U(x) \leqq \sum_N |\langle x, a_n\rangle| \quad \text{for all} \quad x \in E$$

*holds.*

*Proof.* Since the topological dual of the normed space $E(V)$ can be identified with $E'(V^0)$, the inequalities

$$p_U(x) \leqq \sum_N |\langle x, a_n\rangle| \quad \text{for} \quad x \in E$$

and

$$p[x(U)] \leqq \sum_N |\langle x(V), a_n\rangle| \quad \text{for} \quad x(V) \in E(V)$$

are equivalent. The second estimate implies that the canonical mapping from $E(V)$ onto $E(U)$ is quasinuclear if the relation

$$\sum_N p_{V^0}(a_n) < +\infty$$

is valid at the same time. Consequently, properties $(N')$ and $(Q)$ are equivalent for each fundamental system of zero neighborhoods.

**4.1.5.** In the same way we obtain the

**Proposition.** *A locally convex space E is nuclear if and only if some, resp. each, fundamental system $\mathfrak{U}_{\mathfrak{F}}(E)$ of zero neighborhood has the following property:*

*(P) For each zero neighborhood $U \in \mathfrak{U}_{\mathfrak{F}}(E)$ there is a zero neighborhood $V \in \mathfrak{U}_{\mathfrak{F}}(E)$ and a positive Radon measure $\mu$ defined on the weakly compact polar $V^0$ for which the inequality*

$$p_U(x) \leqq \int_{V^0} |\langle x, a\rangle| \, d\mu \quad \text{for all} \quad x \in E$$

*is valid.*

*Proof.* On the basis of Theorem 2.3.3 property $(P)$ is precisely the same as the assertion that the canonical mapping from $E(V)$ onto $E(U)$ to absolutely summing.

**4.1.6.** *If* $\mathfrak{B}_{\mathfrak{F}}(E)$ is a fundamental system of bounded subsets of a locally convex space $E$, then the polar sets $B^0$ with $B \in \mathfrak{B}_{\mathfrak{F}}(E)$ form a fundamental system of zero neighborhoods for the strong topology of the topological dual space $E'$. Thus by Lemma 4.1.1 we have the

**Proposition.** *For a locally convex space $E$ the strong topological dual $E'_b$ is nuclear if and only if there exists in $E$ a fundamental system of bounded subsets with the equivalent properties* (N) *and* (N').

A locally convex space $E$ is called **dual nuclear** if the strong topological dual $E'_b$ is nuclear.

**4.1.7.** As an immediate consequence of 4.1.3 we have the

**Proposition.** *In a dual nuclear locally convex space every fundamental system of bounded subsets has properties* (N) *and* (N').

## 4.2. Summable Families in Nuclear Locally Convex Spaces

**4.2.1.** Let $E$ be an arbitrary locally convex space. In the sequel we shall write

$$\mathfrak{l}_I^1[E] = \mathfrak{l}_I^1\{E\} \quad \text{or} \quad \mathfrak{l}_I^1(E) = \mathfrak{l}_I^1\{E\}\,,$$

if the two linear spaces $\mathfrak{l}_I^1[E]$ or $\mathfrak{l}_I^1(E)$ coincide with $\mathfrak{l}_I^1\{E\}$ without reference to the topology. If the $\boldsymbol{\varepsilon}$-topology coincides with the $\boldsymbol{\pi}$-topology as well we shall use the notation

$$\mathfrak{l}_I^1[E] \equiv \mathfrak{l}_I^1\{E\} \quad \text{or} \quad \mathfrak{l}_I^1(E) \equiv \mathfrak{l}_I^1\{E\}.$$

**4.2.2. Proposition.** *If $E$ is a nuclear locally convex space, then for each index set $I$ we have the identity*

$$\mathfrak{l}_I^1[E] \equiv \mathfrak{l}_I^1(E) \equiv \mathfrak{l}_I^1\{E\}\,.$$

*Proof.* Because of Proposition 4.1.4 there is for each zero neighborhood $U \in \mathfrak{U}(E)$ a zero neighborhood $V \in \mathfrak{U}(E)$ and a sequence of linear forms $a_n \in E'$ with

$$\sum_N p_{V^0}(a_n) \leqq 1$$

for which we have the inequality

$$p_U(x) \leqq \sum_N |\langle x, a_n\rangle| \quad \text{for all} \quad x \in E\,.$$

Since

$$\sum_I |\langle x_i, a_n\rangle| \leqq p_{V^0}(a_n)\, \boldsymbol{\varepsilon}_V\, [x_i, I]$$

is true for all families $[x_i, I] \in \mathfrak{l}_I^1[E]$, we have the inequality

$$\boldsymbol{\pi}_U[x_i, I] = \sum_I p_U(x_i) \leqq \sum_N \sum_I |\langle x_i, a_n\rangle| \leqq \boldsymbol{\varepsilon}_V[x_i, I]\,.$$

Consequently, all weakly summable families in $E$ are also absolutely summable and we have

$$\boldsymbol{l}_I^1[E] = \boldsymbol{l}_I^1(E) = \boldsymbol{l}_I^1\{E\} .$$

We have at the same time proved that the $\boldsymbol{\varepsilon}$-topology is finer than the $\boldsymbol{\pi}$-topology so that the two topologies must coincide.

**4.2.3.** As a converse to the assertion of 4.2.2 we have the

**Proposition.** *Each locally convex space $E$ for which the identity*

$$\boldsymbol{l}_I^1[E] \equiv \boldsymbol{l}_I^1\{E\} \quad (1) \qquad \text{or} \qquad \boldsymbol{l}_I^1(E) \equiv \boldsymbol{l}_I^1\{E\} \quad (2)$$

*holds for some infinite index set $I$, is nuclear.*

*Proof.* In the first place, since the inclusion

$$\boldsymbol{l}_I^1[E] \subset \boldsymbol{l}_I^1(E) \subset \boldsymbol{l}_I^1\{E\}$$

holds for every locally convex space, (2) always follows from (1).

If (2) is valid then the $\boldsymbol{\varepsilon}$-topology coincides with the $\boldsymbol{\pi}$-topology on $\boldsymbol{l}_I^1(E) = \boldsymbol{l}_I^1\{E\}$ so that for each zero neighborhood $U \in \mathfrak{U}(E)$ there is a zero neighborhood $V \in \mathfrak{U}(E)$ for which we have

$$\boldsymbol{\pi}_U[x_i, I] \leqq \boldsymbol{\varepsilon}_V[x_i, I] \quad \text{for} \quad [x_i, I] \in \boldsymbol{l}_I^1(E) .$$

We now consider an arbitrary finite family $[x_n(V), \mathfrak{n}]$ from $E(V)$. If $\mathfrak{i} = \{i_1, \ldots, i_k\}$ is a finite subset of $I$ which has exactly as many elements as the set $\mathfrak{n} = \{n_1, \ldots, n_k\}$, we form the families $[y_i, I]$ belonging to $\boldsymbol{l}_I^1(E)$ with $y_i = o$ for $i \notin \mathfrak{i}$ and $y_{i_h} = x_{n_h}$ for $h = 1, \ldots, k$. Then, because of the identities

$$\sum_{\mathfrak{n}} p[x_n(U)] = \sum_{\mathfrak{n}} p_U(x_n) = \sum_I p_U(y_i) = \boldsymbol{\pi}_U[y_i, I]$$

and

$$\sup\left\{\sum_{\mathfrak{n}} |\langle x_n(V), b\rangle| : b \in V^0\right\} = \sup\left\{\sum_I |\langle y_i, b\rangle| : b \in V^0\right\} = \boldsymbol{\varepsilon}_V[y_i, I]$$

we have the inequality

$$\sum_{\mathfrak{n}} p[x_n(U)] \leqq \sup\left\{\sum_{\mathfrak{n}} |\langle x_n(V), b\rangle| : b \in V^0\right\}.$$

Hence the canonical mapping from $E(V)$ onto $E(U)$ is absolutely summing, and we have shown that the fundamental system $\mathfrak{U}(E)$ has property $(N')$.

**4.2.4.** By combining the results of 4.2.2 and 4.2.3 we obtain the

**Theorem.** *A locally convex space $E$ is nuclear if and only if for some, resp. each, infinite index set $I$ the identity*

$$\boldsymbol{l}_I^1[E] \equiv \boldsymbol{l}_I^1\{E\} \quad \text{or} \quad \boldsymbol{l}_I^1(E) \equiv \boldsymbol{l}_I^1\{E\}$$

*holds.*

**4.2.5.** For a metric locally convex space $E$ the identity

$$\mathfrak{l}_I^1[E] = \mathfrak{l}_I^1\{E\} \quad \text{or} \quad \mathfrak{l}_I^1(E) = \mathfrak{l}_I^1\{E\}$$

implies that the $\varepsilon$-topology is finer than the $\pi$-topology because of Theorem 2.1.3. Consequently, the two topologies coincide and we have

$$\mathfrak{l}_I^1[E] \equiv \mathfrak{l}_I^1\{E\} \quad \text{or} \quad \mathfrak{l}_I^1(E) \equiv \mathfrak{l}_I^1\{E\}\,.$$

Therefore, we have for metric locally convex spaces the following sharpening of Theorem 4.2.4 which also holds for dual metric locally convex spaces because of Theorems 4.2.12 and 4.3.3.

**Theorem.** *A metric or dual metric locally convex space E is nuclear if and only if for some, resp. each, infinite index set I the identity*

$$\mathfrak{l}_I^1[E] = \mathfrak{l}_I^1\{E\} \quad \textit{or} \quad \mathfrak{l}_I^1(E) = \mathfrak{l}_I^1\{E\}$$

*holds.*

**4.2.6.** The question of the generalisation of the previous theorem is posed as a

**Problem.** *What properties must a locally convex space E have so that from the validity of the identity*

$$\mathfrak{l}_I^1[E] = \mathfrak{l}_I^1\{E\} \quad \textit{or} \quad \mathfrak{l}_I^1(E) = \mathfrak{l}_I^1\{E\}$$

*for an infinite index set I it follows that E is nuclear?*

**4.2.7.** We now extend the previous results of this section to dual nuclear spaces. For this we use the

**Lemma.** *If E is a dual nuclear locally convex space, then for each bounded subset* $\mathbf{B}$ *of* $\mathfrak{l}_I^1[E]$ *there is a bounded subset* $B \in \mathfrak{B}(E)$ *with*

$$\sum_I p_B(x_i) \leqq 1 \quad \textit{for all} \quad [x_i, I] \in \mathbf{B}\,.$$

*Proof.* By hypothesis the number

$$\sigma_U = \sup\{\varepsilon_U[x_i, I] : [x_i, I] \in \mathbf{B}\}$$

is bounded for each zero neighborhood $U \in \mathfrak{U}(E)$, and we can form the set

$$A = \{\sigma_U\, U : U \in \mathfrak{U}(E)\}$$

which belongs to $\mathfrak{B}(E)$. Consequently, we have

$$\sum_{\mathfrak{i}} \lambda_i\, x_i \in A \quad \text{for} \quad [x_i, I] \in \mathbf{B}\,, \qquad |\lambda_i| \leqq 1 \quad \text{and} \quad \mathfrak{i} \in \mathfrak{F}(I)\,,$$

and the inequality

$$\varepsilon_A[x_i, I] \leqq 1 \quad \text{for all} \quad [x_i, I] \in \mathbf{B}$$

holds.

We now determine a set $B \in \mathfrak{B}(E)$ with $A < B$ such that the canonical mapping $E(A, B)$ is absolutely summing. Since we can always make the $\boldsymbol{\pi}$-norm of $E(A, B)$ smaller than 1, the inequality

$$\boldsymbol{\pi}_B[x_i, I] \leqq \boldsymbol{\varepsilon}_A[x_i, I] \quad \text{for all} \quad [x_i, I] \in \boldsymbol{l}_I^1[E(A)]$$

is valid and the estimate

$$\sum_I p_B(x_i) \leqq 1 \quad \text{for all} \quad [x_i, I] \in \mathbf{B}$$

is proved.

**4.2.8. Proposition.** *If $E$ is a dual nuclear locally convex space, then for each index set $I$ the identity*

$$\boldsymbol{l}_I^1[E] = \boldsymbol{l}_I^1(E) = \boldsymbol{l}_I^1\{E\} = \boldsymbol{l}_I^1\langle E\rangle$$

*holds.*

*Proof.* Since each weakly summable family $[x_i, I]$ from $E$ can be regarded as a bounded singleton from $\boldsymbol{l}_I^1[E]$ there is by the lemma just proved a set $B \in \mathfrak{B}(E)$ with

$$\sum_I p_B(x_i) \leqq 1 \, .$$

Consequently, all weakly summable families from $E$ are also totally summable and we have

$$\boldsymbol{l}_I^1[E] = \boldsymbol{l}_I^1(E) = \boldsymbol{l}_I^1\{E\} = \boldsymbol{l}_I^1\langle E\rangle \, .$$

**4.2.9. Proposition.** *Each dual nuclear space $E$ has property* $(B)$.

*Proof.* Since each bounded subset $\mathbf{B}$ of $\boldsymbol{l}_I^1\{E\}$ can also be considered as a bounded subset of $\boldsymbol{l}_I^1[E]$, there exists by Lemma 4.2.7 a bounded subset $B \in \mathfrak{B}(E)$ with

$$\sum_I p_B(x_i) \leqq 1 \quad \text{for all} \quad [x_i, I] \in \mathbf{B} \, .$$

Consequently the locally convex space has property $(B)$.

**4.2.10.** As a converse of the statements of 4.2.8 and 4.2.9 we have the

**Proposition.** *Each locally convex space $E$ with property $(B)$ for which the identity*

$$\boldsymbol{l}_I^1[E] = \boldsymbol{l}_I^1\{E\} \quad (1) \qquad \textit{or} \qquad \boldsymbol{l}_I^1(E) = \boldsymbol{l}_I^1\{E\} \quad (2)$$

*holds for some infinite index set $I$ is dual nuclear.*

*Proof.* To begin with, we recall that since the inclusion

$$\boldsymbol{l}_I^1\{E\} \subset \boldsymbol{l}_I^1(E) \subset \boldsymbol{l}_I^1[E]$$

is true for every locally convex space $E$, (2) always follows from (1).

For an arbitrary set $A \in \mathfrak{B}(E)$,

$$\mathbf{B} = \{[x_n, N] \in \boldsymbol{l}_N^1[E(A)] : \boldsymbol{\varepsilon}_A[x_n, N] \leqq 1\}$$

is a bounded subset of $\boldsymbol{l}_N^1[E]$. Since from (2) it follows that the identity mapping from $E$ into itself is absolutely summing, $\mathbf{B}$ must by Theorem 2.1.2, be a bounded subset of $\boldsymbol{l}_N^1\{E\}$ as well. Therefore, there exists a set $B \in \mathfrak{B}(E)$ with

$$\sum_N p_B(x_n) \leqq 1 \quad \text{for } [x_n, N] \in \mathbf{B}.$$

From this statement it follows that the inequality

$$\boldsymbol{\pi}_B[x_n, N] \leqq \boldsymbol{\varepsilon}_A[x_n, N] \quad \text{for all } [x_n, N] \in \boldsymbol{l}_I^1[E(A)]$$

is valid. Consequently, the canonical mapping from $E(A)$ into $E(B)$ is absolutely summing and we have shown that the fundamental system $\mathfrak{B}(E)$ has property $(N)$.

**4.2.11.** By combining the results of 4.2.8, 4.2.9 and 4.2.10 we obtain the

**Theorem.** *A locally convex space E is dual nuclear if and only if it has property (B) and for some, resp. each, infinite index set I the identity*

$$\boldsymbol{l}_I^1[E] = \boldsymbol{l}_I^1\{E\} \quad \text{or} \quad \boldsymbol{l}_I^1(E) = \boldsymbol{l}_I^1\{E\}$$

*holds.*

**4.2.12.** As a special case we obtain by means of 1.5.8 the following counterpart to Theorem 4.2.5.

**Theorem.** *A metric or dual metric locally convex space E is dual nuclear if and only if for some, resp. each, infinite index set I the identity*

$$\boldsymbol{l}_I^1[E] = \boldsymbol{l}_I^1\{E\} \quad \text{or} \quad \boldsymbol{l}_I^1(E) = \boldsymbol{l}_I^1\{E\}$$

*is valid.*

## 4.3. The Topological Dual of Nuclear Locally Convex Spaces

**4.3.1.** We now give a necessary and sufficient condition that the strong topological dual $E'_b$ of a nuclear locally convex space $E$ is also nuclear.

**Theorem.** *A nuclear locally convex space E is dual nuclear if and only if it has property (B).*

*Proof.* The necessity of our condition follows immediately from 4.2.9. To show the sufficiency we note that for each nuclear locally convex space $E$ the identity

$$\boldsymbol{l}_I^1(E) = \boldsymbol{l}_I^1\{E\}$$

is valid. If the locally convex space $E$ has property $(B)$ as well, it must be dual nuclear by 4.2.10.

**4.3.2.** On the other hand, if we know that the strong topological dual $E'_b$ of a locally convex space $E$ is nuclear, then we can ask under what assumptions must $E$ also be nuclear. The statement of this problem is meaningful only if the topology of $E$, which in general is not uniquely determined by $E_b'$, is fixed by certain additional assumptions. In particular, this is the case for quasi-barrelled locally convex spaces.

As a preliminary to the following theorem we first prove the

**Lemma.** *A quasi-barrelled locally convex space $E$ is nuclear if and only if $E_b'$ is dual nuclear.*

*Proof.* Since $E$ is, by hypothesis, quasi-barrelled,

$$\mathfrak{U}^0(E_b') = \{U^0 : U \in \mathfrak{U}(E)\}$$

is a fundamental system of bounded subsets in $E_b'$. Our conclusion now follows from the fact that $\mathfrak{U}(E)$ has property $(N)$ as a fundamental system of zero neighborhoods if and only if $\mathfrak{U}^0(E_b')$ has property $(N)$ as a fundamental system of bounded subsets.

We now obtain the

**Theorem.** *A dual nuclear quasi-barrelled locally convex space $E$ is nuclear if and only if its strong topological dual $E_b'$ has property $(B)$.*

*Proof.* Since $E_b'$ is nuclear by hypothesis, our conclusion follows from the preceeding lemma and Theorem 4.3.1.

**4.3.3.** In order to include dual nuclear locally convex spaces in our investigation we must improve the previous lemma.

**Lemma.** *A $\sigma$-quasi-barrelled locally convex space $E$ is nuclear if $E_b'$ is dual nuclear.*

*Proof.* We consider an arbitrary zero neighborhood $U \in \mathfrak{U}(E)$. Then $U^0$ belongs to $\mathfrak{B}(E_b')$ and by hypothesis there is a set $B \in \mathfrak{B}(E_b')$ with $U^0 < B$, such that the canonical mapping from $E'(U^0)$ into $E'(B)$ is nuclear. Consequently, the identity

$$a = \sum_N \langle \mathfrak{x}_n, a \rangle\, b_n \quad \text{for all } a \in E'(U^0)$$

is valid in $E'(B)$. Here we can assume that the linear forms $b_n$ lie in $B$, and the norms $p(\mathfrak{x}_n)$ of the linear forms $\mathfrak{x}_n$ defined on $E'(U^0)$ satisfy the inequality

$$\sum_N p(\mathfrak{x}_n) < +\infty .$$

Since the set of linear forms $b_n$ is countable and strongly bounded, there is, by hypothesis, a zero neighborhood $V \in \mathfrak{U}(E)$ with $b_n \subset V^0$. We now see that the canonical mapping of $E'(U^0)$ into $E'(V^0)$ is nuclear because we have

$$a = \sum_N \langle \mathfrak{x}_n, a \rangle\, b_n \quad \text{for all } a \in E'(U^0)$$

and

$$\sum_N p(\mathfrak{x}_n)\, p_{V^\circ}(b_n) < +\infty .$$

We have thus shown that the fundamental system $\mathfrak{U}(E)$ has property $(N)$.

As a special result we obtain the following

**Theorem.** *A metric or dual metric locally convex space E is nuclear if and only if it is dual nuclear.*

*Proof.* Since each metric or dual metric locally convex space has property $(B)$ by Theorem 1.5.8, it follows from Theorem 4.3.1 that all nuclear metric or dual metric locally convex spaces are also dual nuclear.

On the other hand, if we have a dual nuclear metric, resp. dual metric, locally convex space $E$, then the strong topological dual $E'_b$ is also dual nuclear because it is a dual metric, resp. metric, locally convex space. From the previously established lemma it follows that $E$ must be nuclear.

Because of the great significance of the theorem just proven we now give a second version.

**Theorem′.** *A metric or dual metric locally convex space E has a nuclear strong topological dual $E'_b$ if and only if it is itself nuclear.*

**4.3.4.** As an illustration of our result we consider the locally convex space $\Omega_I$ already introduced in 1.5.7. Since the normed spaces $\Omega_I(U)$ are finite dimensional for all zero neighborhoods $U \in \mathfrak{U}(\Omega_I)$, $\Omega_I$ is nuclear. On the other hand, in the case of an uncountable index set $I$, the locally convex space $\Omega_I$ cannot be dual nuclear, because it does not have property $(B)$.

The topological dual of $\Omega_I$ can be identified with the linear space $\Phi_I$ of all families of numbers $[\alpha_i, I]$, which contain only finitely many $\alpha_i \neq 0$. We here obtain the strong topology of $\Phi_I$ from the system of all semi-norms

$$p[\alpha_i, I] = \sum_I |\alpha_i \xi_i| \quad \text{with} \quad [\xi_i, I] \in \Omega_I .$$

Since we can also consider $\Omega_I$ as the strong topological dual of $\Phi_I$, the locally convex spaces $\Omega_I$ and $\Phi_I$ are reflexive. Consequently $\Phi_I$ for an uncountable index set is a dual nuclear but not a nuclear locally convex space.

**Proposition.** *There are nuclear, resp. non-nuclear, quasi-barrelled locally convex spaces which have a non-nuclear, resp. nuclear, strong topological dual.*

## 4.4. Properties of Nuclear Locally Convex Spaces

**4.4.1.** We first show that many questions about nuclear locally convex spaces can be reduced to questions about Hilbert spaces. In fact, we have the

**Proposition.** *In each nuclear locally convex space E there is a fundamental system $\mathfrak{U}_{\mathfrak{H}}(E)$ of zero neighborhoods for which all Banach spaces $\widetilde{E(U)}$ and $E'(U^0)$ are Hilbert spaces.*

*Proof.* We denote by $\mathfrak{U}_{\mathfrak{H}}(E)$ the collection of all zero neighborhoods $W \in \mathfrak{U}(E)$ for which $\widetilde{E(W)}$ and thus $E'(W^0)$ is a Hilbert space, and show that for each zero neighborhood $U \in \mathfrak{U}(E)$ there is a zero neighborhood $W \in \mathfrak{U}_{\mathfrak{H}}(E)$ with $W \subset U$.

For this purpose we determine by means of Proposition 4.1.5 a zero neighborhood $V \in \mathfrak{U}(E)$ such that there exists on $V^0$ a positive Radon measure $\mu$ with

$$p_U(x) \leqq \int_{V^0} |\langle x, a\rangle| \, d\mu \qquad \text{for all} \quad x \in E .$$

Then from the expression

$$p_W(x) = \left\{\mu(V^0) \int_{V^0} |\langle x, a\rangle|^2 \, d\mu\right\}^{1/2}$$

we obtain on $E$ a continuous semi-norm with $p_W(x) \leqq \mu(V^0)\, p_V(x)$ which can be derived from the semi-scalar product

$$(x, y)_W = \mu(V^0) \int_{V^0} \langle x, a\rangle \overline{\langle y, a\rangle} \, d\mu .$$

Therefore, $\widetilde{E(W)}$ is a Hilbert space. Moreover, an application of the Hölder inequality shows that $p_U(x) \leqq p_W(x)$. Consequently, we have $W \subset U$, and our assertion is proved.

**4.4.2.** As a simple consequence of the proposition just proved we have the

**Theorem.** *A locally convex space E is nuclear if and only if it contains a fundamental system $\mathfrak{U}_{\mathfrak{H}}(E)$ of zero neighborhoods with the following two properties*

*($H_1$) For each zero neighborhood $U \in \mathfrak{U}_{\mathfrak{H}}(E)$, $E'(U^0)$ is a Hilbert space.*

*($H_2$) For each zero neighborhood $U \in \mathfrak{U}_{\mathfrak{H}}(E)$ there is a zero neighborhood $V \in \mathfrak{U}_{\mathfrak{H}}(E)$ with $V < U$ such that $E'(U^0, V^0)$ is a Hilbert-Schmidt-mapping.*

*Proof.* Since in Hilbert spaces the absolutely summing mappings coincide with the Hilbert-Schmidt-mappings by Theorem 2.5.5, property ($H_2$) is equivalent to property ($N$) given in 4.1.2. If there exists in a locally convex space $E$ a fundamental system of zero neighborhoods with properties $(H_1)$ and $(H_2)$, then the space must be nuclear.

On the other hand, if we consider an arbitrary nuclear locally convex space $E$ there is by 4.4.1 a fundamental system $\mathfrak{U}_{\mathfrak{H}}(E)$ of zero neighborhoods with property $(H_1)$. Since $E$ is supposed to be nuclear, $\mathfrak{U}_{\mathfrak{H}}(E)$ has property $(N)$ which is equivalent to property $(H_2)$.

**4.4.3. Proposition.** *For a nuclear or dual nuclear locally convex space $E$ all canonical mappings $E(A, U)$ with $A \in \mathfrak{B}(E)$ and $U \in \mathfrak{U}(E)$ are nuclear.*

*Proof.* Our assertion results from the fact that we can write the mapping $E(A, U)$ in the form

$$E(A, U) = E(V, U)\, E(A, V) \quad \text{resp.} \quad E(A, U) = E(B, U)\, E(A, B).$$

Here the zero neighborhood $V \in \mathfrak{U}(E)$ with $V < U$, resp. subset $B \in \mathfrak{B}(E)$ with $A < B$, can be so chosen that the canonical mapping $E(V, U)$, resp. $E(A, B)$, is nuclear.

**4.4.4.** In general, the nuclearity of all canonical mappings $E(A, U)$ with $A \in \mathfrak{B}(E)$ and $U \in \mathfrak{U}(E)$ is not sufficient for the nuclearity or dual nuclearity of a locally convex space $E$. However, we have the

**Proposition.** *If in the locally convex space $E$ every canonical mapping $E(K, U)$ with $K \in \mathfrak{K}(E)$ and $U \in \mathfrak{U}(E)$ is absolutely summing we then have the identity*

$$\mathfrak{l}_I^1(E) = \mathfrak{l}_I^1\{E\}$$

*for each index set $I$.*

*Proof.* For an arbitrary summable family $[x_i, I]$ from $E$ we set

$$H_{\mathfrak{i}} = \left\{\sum_{\mathfrak{i}} \alpha_i x_i : |\alpha_i| \leqq 1\right\} \quad \text{and} \quad H = \bigcup \{H_{\mathfrak{i}} : \mathfrak{i} \in \mathfrak{F}(I)\}\,.$$

Since for each zero neighborhood $V \in \mathfrak{U}(E)$ there is a set $\mathfrak{i}_0 \in \mathfrak{F}(I)$ with

$$\sum_{\mathfrak{j}} |\langle x_i, a\rangle| \leqq 1 \quad \text{for } a \in V^0 \quad \text{and} \quad \mathfrak{j} \in \mathfrak{F}(I) \text{ with } \mathfrak{j} \cap \mathfrak{i}_0 = \emptyset\,,$$

we have for all elements

$$x = \sum_{\mathfrak{i}} \alpha_i x_i \in H$$

the statement

$$x = \sum_{\mathfrak{i} \cap \mathfrak{i}_0} \alpha_i x_i + \sum_{\mathfrak{i} \setminus \mathfrak{i}_0} \alpha_i x_i \in H_{\mathfrak{i}_0} + V\,.$$

Hence we get the relation

$$H \subset H_{\mathfrak{i}_0} + V\,.$$

Since the finite dimensional and bounded set $H_{i_0}$ is precompact, there exist finitely many elements $y_1, \ldots, y_s \in E$ with

$$H_{i_0} \subset \bigcup_{n=1}^{s} \{y_n + V\} .$$

This yields the expression

$$H \subset \bigcup_{n=1}^{s} \{y_n + 2V\} ,$$

and we have shown that $H$ is a precompact subset of $E$. We can hence find a set $K$ in $\mathfrak{K}(E)$ with $H \subset K$.

Since

$$\sum_{i} \alpha_i x_i \in K \quad \text{for } i \in \mathfrak{F}(I) \quad \text{and} \quad |\alpha_i| \leqq 1 ,$$

the inequality

$$\sum_I |\langle x_i, \mathfrak{a}\rangle| \leqq p'_K(\mathfrak{a}) < +\infty$$

holds for all linear forms $\mathfrak{a} \in [E(K)]'$. Consequently, the family $[x_i, I]$ belongs to $\mathfrak{l}^1_I[E(K)]$. Since, however, all canonical mappings from $E(K)$ into $E(U)$ with $U \in \mathfrak{U}(E)$ are assumed to be absolutely summing, we have

$$\sum_I p_U(x_i) = \sum_I p[x_i(U)] < +\infty \qquad \text{for all } U \in \mathfrak{U}(E) .$$

We have thus shown that under the given hypothesis every summable family from $E$ is also absolutely summable. The identity

$$\mathfrak{l}^1_I(E) = \mathfrak{l}^1_I[E]$$

thus holds for each index set.

**4.4.5.** By combining the previous result with the statement of Theorem 4.2.10 we obtain the

**Theorem.** *A locally convex space $E$ with property (B) is dual nuclear if and only if in it every canonical mapping $E(K, U)$ with $K \in \mathfrak{K}(E)$ and $U \in \mathfrak{U}(E)$ is nuclear, resp. quasinuclear, resp. absolutely summing.*

**4.4.6.** Moreover, by 4.3.3 we have the

**Theorem.** *A metric or dual metric locally convex space $E$ is nuclear if and only if in it every canonical mapping $E(K, U)$ with $K \in \mathfrak{K}(E)$ and $U \in \mathfrak{U}(E)$ is nuclear, resp. quasinuclear, resp. absolutely summing.*

**4.4.7.** As a simple consequence of 4.3.3 we obtain the

**Proposition.** *In each nuclear or dual nuclear locally convex space $E$ all bounded subsets are precompact.*

*Proof.* Since each arbitrary bounded subset of $E$ is contained in a closed and absolutely convex bounded subset, it suffices to prove our assertion for all sets $A \in \mathfrak{B}(E)$. But now the canonical mapping from $E(A)$ into $E(U)$ is nuclear and thus precompact for each zero neighborhood $U \in \mathfrak{U}(E)$. Consequently,

$$A(U) = \{x(U) \in E(U) : x \in A\}$$

must be a precompact subset of $E(U)$. We can therefore determine finitely many elements $x_1, \ldots, x_s \in E$ such that the relation

$$A(U) \subset \bigcup_{n=1}^{s} \{x_n(U) + U(U)\} \tag{1}$$

holds. Here

$$U(U) = \{x(U) \in E(U) : x \in U\}$$

is the closed unit ball of $E(U)$. By (1) we also have

$$A \subset \bigcup_{n=1}^{s} \{x_n + U\} . \tag{2}$$

Since $U$ was an arbitrary zero neighborhood from $\mathfrak{U}(E)$ our assertion is proven.

**4.4.8.** As a supplement to the statement of 4.4.7 we get the

**Proposition.** *In each dual nuclear locally convex space $E$ all bounded subsets are separable.*

*Proof.* Each arbitrary bounded subset $A_0$ of $E$ is contained in a set $A \in \mathfrak{B}(E)$. We now determine a set $B \in \mathfrak{B}(E)$ with $A < B$ such that the canonical mapping from $E(A)$ into $E(B)$ is nuclear. Then $A_0$ is a precompact and hence separable subset of the normed space $E(B)$. Since the identity mapping from $E(B)$ into $E$ is continuous, $A_0$ has the same property as a subset of $E$.

**4.4.9.** In a nuclear locally convex space $E$ there may be bounded subsets which are not separable. In contrast to this we have the

**Proposition.** *For each nuclear locally convex space $E$, every normed space $E(U)$ with $U \in \mathfrak{U}(E)$ is separable.*

*Proof.* We determine for an arbitrary given zero neighborhood $U \in \mathfrak{U}(E)$ a zero neighborhood $V \in \mathfrak{U}(E)$ with $V < U$, such that the canonical mapping from $E(V)$ onto $E(U)$ is nuclear. Then by 3.1.6 the space $E(U)$ must be separable because it is the range of a nuclear mapping.

**4.4.10.** We now show that there are no arbitrarily large nuclear metric or dual metric locally convex spaces. In fact, we have the

**Theorem.** *All nuclear metric or dual metric locally convex spaces are separable.*

*Proof.*

(1) We first treat a nuclear metric locally convex space $E$, in which we determine a countable fundamental system

$$\mathfrak{U}_{\mathfrak{F}}(E) = \{U_n : n = 1, 2, \ldots\}$$

of zero neighborhoods. Then there is for each natural number $n$ a sequence of elements $x_{mn} \in E$ such that the classes $x_{mn}(U_n)$ with $m = 1, 2, \ldots$ are dense in the separable normed space $E(U_n)$.

If $x$ is an arbitrary element from $E$, then there exist for each zero neighborhood $U \in \mathfrak{U}(E)$ a zero neighborhood $U_n$ with $U_n \subset U$ and an element $x_{mn}$ with

$$p_U(x - x_{mn}) \leqq p_{U_n}(x - x_{mn}) = p\,[x(U_n) - x_{mn}(U_n)] \leqq 1\,.$$

This proves that the double sequence of elements $x_{mn}$ is dense in $E$.

(2) We now treat a nuclear dual metric locally convex space $E$, in which we determine a countable fundamental system

$$\mathfrak{B}_{\mathfrak{F}}(E) = \{B_n : n = 1, 2, \ldots\}$$

of bounded subsets. Since $E$ is also dual nuclear, every set $B_n$ is separable. Thus for each natural number $n$ there is a sequence of elements $x_{mn}$ with $m = 1, 2, \ldots$ which are dense in $B_n$. Since each element $x \in E$ is contained in at least one set $B_n$, we see that the double sequence $x_{mn}$ is dense in $E$.

**4.4.11. Proposition.** *Each quasi-complete nuclear or dual nuclear locally convex space $E$ is semi-reflexive.*

*Proof.* Since in a nuclear or dual nuclear locally convex space all bounded subsets are precompact by 4.4.7, all closed and bounded subsets of $E$ must be compact by Proposition 0.5.7 and hence weakly compact. Our conclusion is now obtained from 0.7.2.

**4.4.12.** To prove the following theorem we need the

**Lemma.** *Each nuclear dual metric locally convex space $E$ is quasi-barrelled.*

*Proof.* Since the metric locally convex space $E'_b$ is dual nuclear, its bounded subsets are separable by 4.4.8. Thus, each bounded subset $B$ of $E'$ lies in the closed hull of a sequence of linear forms $a_n \in B$. Since $E$ as a dual metric locally convex space is $\sigma$-quasi-barrelled, there is a zero neighborhood $U \in \mathfrak{U}(E)$ with $a_n \in U^0$. Consequently, the entire set $B$ lies in $U^0$. Thus our assertion is proved.

On the basis of the definition given in 0.7.3 we now obtain the

**Theorem.** *All nuclear $(F)$- and $(F')$- spaces are reflexive.*

**4.4.13.** By combining the results of 4.3.3 and 4.4.12 we obtain the noteworthy

**Theorem.** *The correspondence* $E \leftrightarrow F$ *with*

$$F = E'_b \quad \text{and} \quad E = F'_b$$

*establishes a one-to-one relation between the nuclear* $(F)$*-spaces and the nuclear* $(F')$*-spaces.*

**4.4.14.** Having previously treated nuclear metric and dual metric locally convex spaces we finally characterize nuclear normed spaces.

**Theorem.** *Only the finite dimensional normed spaces are nuclear.*

*Proof.* In each normed space $E$ the closed unit ball $U$ is bounded. If $E$ is assumed to be nuclear, then $U$ must be precompact as well by Proposition 4.4.7. Thus it follows from Proposition 0.8.3 that $E$ is finite dimensional.

On the other hand, for a finite dimensional locally convex space $E$, the canonical mapping from $E(U)$ onto $E(U)$ is finite and hence nuclear for each zero neighborhood $U \in \mathfrak{U}(E)$. Consequently, the fundamental system $\mathfrak{U}(E)$ has property $(N')$.

Chapter 5

# Permanence Properties of Nuclearity

There are many possible ways of obtaining new locally convex spaces from a given set of locally convex spaces. In this chapter we show that in most cases nuclearity carries over from the original spaces to the spaces newly produced. The following cases are treated: linear subspaces and quotient spaces, topological products and sums, locally convex kernels and hulls, complete and quasi-complete hulls, locally convex tensor products and spaces of continuous linear mappings. Almost all of the propositions in this chapter were first proved by A. Grothendieck [3].

## 5.1. Subspaces and Quotient Spaces

**5.1.1. Proposition.** *Each linear subspace $F$ of a nuclear locally convex space $E$ is also nuclear.*

*Proof.* First we note that the sets $U_0 = F \cap U$ with $U \in \mathfrak{U}(E)$ form a fundamental system of zero neighborhoods in $F$. Moreover, from the relation

$$p[x(U_0)] = p_{U_0}(x) = p_U(x) = p[x(U)] \quad \text{with } x \in F,$$

it follows that we can identify the normed space $F(U_0)$ with the linear subspace of $E(U)$ which consists of all equivalence classes $x(U)$ with $x \in F$.

Since the fundamental system $\mathfrak{U}(E)$ has property $(N')$, there is for each zero neighborhood $U \in \mathfrak{U}(E)$ a zero neighborhood $V \in \mathfrak{U}(E)$ with $V < U$ such that the canonical mapping $E(V, U)$ is quasinuclear. Then the canonical mapping $F(V_0, U_0)$ as the restriction of $E(V, U)$ to $F(V_0)$ must also be quasinuclear. Consequently, the fundamental system of all zero neighborhoods $U_0 = F \cap U$ has property $(N')$.

**5.1.2. Proposition.** *Each linear subspace $F$ of a dual nuclear locally convex space $E$ is also dual nuclear.*

*Proof.* We first note that the sets $A_0 = F \cap A$ with $A \in \mathfrak{B}(E)$ form a fundamental system of bounded subsets in $F$. Moreover, from the relation

$$p_{A_0}(x) = p_A(x) \quad \text{with } x \in F(A_0),$$

it follows that we can identify the normed space $F(A_0)$ with the linear subspace $F \cap E(A)$ of $E(A)$.

Since the fundamental system $\mathfrak{B}(E)$ has property $(N)$, there is for each set $A \in \mathfrak{B}(E)$ a set $B \in \mathfrak{B}(E)$ with $A < B$ such that the canonical mapping $E(A, B)$ is quasinuclear. Then the canonical mapping $F(A_0, B_0)$ as the restriction of $E(A, B)$ to $F(A_0)$ must be quasinuclear. Consequently, the fundamental system of all bounded subsets $A_0 = F \cap A$ has property $(N)$.

**5.1.3. Proposition.** *Each quotient space $Q = E/F$ of a nuclear locally convex space $E$ by a closed linear subspace $F$ is also nuclear.*

*Proof.* For each linear form $a \in F^0$ a linear form $\mathfrak{a}$ on the quotient space $Q$ is given by

$$\langle x(F), \mathfrak{a}\rangle = \langle x, a\rangle \quad \text{for } x \in E .$$

Since we can obtain all linear forms $\mathfrak{a} \in Q'$ in this way, the topological dual of $Q$ can be identified with the linear subspace $F^0$ of $E'$.

We now note that the sets

$$\overline{U(F)} = \{x(F) \in Q : \alpha\, x \in U + F \quad \text{for all } \alpha \in K \quad \text{with } |\alpha| < 1\}$$

form a fundamental system of zero neighborhoods in $Q$, as $U$ ranges over all zero neighborhoods in $\mathfrak{U}(E)$. Moreover, from the relation

$$F^0 \cap U^0 = \overline{U(F)}^0 ,$$

it follows that the normed space $Q'\left(\overline{U(F)}^0\right)$ coincides with the linear subspace $F^0 \cap E'(U^0)$ of $E'(U^0)$. Here we identify the linear spaces $Q'$ and $F^0$ in the way presented above.

Since the fundamental system $\mathfrak{U}(E)$ has property $(N)$, there is for each zero neighborhood $U \in \mathfrak{U}(E)$ a zero neighborhood $V \in \mathfrak{U}(E)$ with $V < U$ such that the canonical mapping $E'(U^0, V^0)$ is quasinuclear. Then the canonical mapping $Q'\left(\overline{U(F)}^0, \overline{V(F)}^0\right)$ as the restriction of $E'(U^0, V^0)$ to $F^0 \cap E'(U^0)$ must be quasinuclear. Consequently, the fundamental system of all zero neighborhoods $\overline{U(F)}$ has property $(N)$.

**5.1.4. Problem.** *Is every quotient space $Q = F/E$ of a dual nuclear locally convex space $E$ by a closed linear subspace $F$ also dual nuclear?*

*Remark.* The answer to this question is certainly positive if each bounded subset of $Q$ is the image of a bounded subset of $E$. In particular this assumption is satisfied for dual nuclear $(F)$-spaces.

**Lemma.** *If $F$ is a closed linear subspace of the dual nuclear $(F)$-space $E$, then the sets*

$$B(F) = \{x(F) \in Q : x \in B\} \quad \textit{with } B \in \mathfrak{B}(E)$$

*form a fundamental system of bounded subsets of the quotient space* $Q = E/F$.

*Proof.* Since $Q$ is nuclear it follows from Theorem 9.5.2 that the sets

$$A[x_r(F), R] = \left\{x(F) \in Q : x(F) = \sum_R \xi_r x_r(F) \quad \text{with} \quad \sum_R |\xi_r| \leqq 1\right\}$$

form a fundamental system of bounded subsets in $Q$ as $[x_r(F), R]$ ranges over all totally summable sequences.

We now consider in $E$ a countable fundamental system of zero neighborhoods $U_n$ with $U_1 \supset U_2 \supset \cdots$. Then the sets

$$\mathfrak{r}_n = \{r \in R : p_{U_n^\sim}[x_r(F)] \geqq 1\},$$

are finite for each totally summable sequence $[x_r(F), R]$ and we have

$$\mathfrak{r}_1 \subset \mathfrak{r}_2 \subset \cdots.$$

We now set

$$y_r = 0 \quad \text{for} \quad r \notin \bigcup_N \mathfrak{r}_n \quad \text{and} \quad y_r = x_r \quad \text{for} \quad r \in \mathfrak{r}_1,$$

and in the case $r \in \mathfrak{r}_{n+1} \backslash \mathfrak{r}_n$ we determine the elements $y_r$ such that

$$x_r - y_r \in F \quad \text{and} \quad p_{U_n}(y_r) < 1.$$

We then have

$$y_r(F) = x_r(F) \quad \text{for all} \quad r \in R.$$

In fact if the number $r \in R$ lies in no set $\mathfrak{r}_n$ we then get the statement

$$p_{U_n}[x_r(F)] \leqq 1 \quad \text{for all} \quad n \in N.$$

But this is possible only if we also have $x_r(F) = o(F)$.

For each number $r \notin \mathfrak{r}_m$ it is either true that $r \notin \bigcup_N \mathfrak{r}_n$ and hence $p_{U_m}[x_r(F)] = 0$ or there is a natural number $n$ with $n \geqq m$ and $r \in \mathfrak{r}_{n+1} \backslash \mathfrak{r}_n$ so that the inequality

$$p_{U_m}(y_r) \leqq p_{U_n}(y_r) < 1$$

holds. Consequently we have

$$p_{U_m}(y_r) < 1 \quad \text{for} \quad r \notin \mathfrak{r}_m.$$

Thus $[y_r, R]$ is a null sequence for which we can determine a set $B \in \mathfrak{B}(E)$ with $y_r \in B$ for $r \in R$. Therefore, our assertion is proved because we have

$$A_r[x(F), R] \subset B(F).$$

**5.1.5** From 5.1.1 we immediately obtain the

**Proposition.** *Each closed linear subspace $F$ of a nuclear $(F)$-space $E$ is also a nuclear $(F)$-space.*

**5.1.6.** Since all quotient spaces of $(F)$-spaces are also $(F)$-spaces we obtain from 5.1.3 the

**Proposition.** *Each quotient space $Q = E/F$ of a nuclear $(F)$-space $E$ by a closed linear subspace $F$ is also a nuclear $(F)$-space.*

*Remark.* The completeness of $Q$ also results from the fact, proved in Lemma 5.1.4, that each Cauchy-sequence $[x_n(F), N]$ is contained in the compact image $B(F)$ of a compact subset $B$ of $E$.

**5.1.7.** It can be shown that for several $(F')$-spaces there exist closed linear subspaces which are not $(F')$-spaces. For nuclear spaces, however, we have the

**Proposition.** *Each closed linear subspace $F$ of a nuclear $(F')$-space $E$ is also a nuclear $(F')$-space.*

*Proof.* Our conclusion results from the fact that the linear subspace $F$ can be identified with the strong topological dual of the nuclear $(F)$-space $E'_b/F^0$. For this purpose we associate with each element $x \in F$ the linear form $\mathfrak{x}$ on $E'_b/F^0$ defined by

$$\langle \mathfrak{x}, a(F^0) \rangle = \langle x, a \rangle .$$

We obtain all continuous linear forms on $E'_b/F^0$ in this way because $E$ is reflexive. Finally, by Lemma 5.1.4, the strong topology of $F = (E'_b/F^0)'_b$ can be obtained from the semi-norms

$$p'_B(x) = \sup \{|\langle \mathfrak{x}, a(F^0) \rangle| : a \in B\} \quad \text{with} \quad B \in \mathfrak{B}(E'_b) .$$

It thus coincides with the topology induced by $E$.

**5.1.8.** There exist examples of $(F')$-spaces which have non-complete quotient spaces. In the case of nuclear $(F')$-spaces, however, we have the

**Proposition.** *Each quotient space $Q = E/F$ of a nuclear $(F')$-space $E$ by a closed linear subspace $F$ is also a nuclear $(F')$-space.*

*Proof.* This results from the fact that the quotient space $Q$ can be identified with the strong topological dual of the nuclear $(F)$-space $F^0$. For this purpose we associate with each equivalence class $x(F) \in Q$ the linear form $\mathfrak{x}$ defined on $F^0$ by

$$\langle \mathfrak{x}, a \rangle = \langle x, a \rangle .$$

We obtain all continuous linear forms on $F^0$ in this way because of Hahn Banach Theorem and the fact that $E$ is reflexive. Since the sets $U^0 \cap F^0$ with $U \in \mathfrak{U}(E)$ form a fundamental system of bounded subsets in $F^0$, we obtain the strong topology on $Q = (F^0)'_b$ from the semi-norms

$$q_U[x(F)] = \sup \{|\langle x, a \rangle| : a \in U^0 \cap F^0\} \quad \text{with} \quad U \in \mathfrak{U}(E) ,$$

and, as is well-known, the quotient space topology on $Q$ from the semi-norms

$$p_U[x(F)] = \inf \{p_U(x+y) : y \in F\} \quad \text{with} \quad U \in \mathfrak{U}(E)\,.$$

The assertion that the two topologies coincide results from the identity

$$p_U[x(F)] = q_U[x(F)]\,,$$

which we now prove.

From the inequality

$$|\langle x, a\rangle| = |\langle x+y, a\rangle| \leqq p_U(x+y)$$

valid for $x \in E$, $y \in F$ and $a \in U^0 \cap F^0$ we obtain the estimate

$$q_U[x(F)] \leqq p_U(x+y) \quad \text{for all} \quad y \in F\,.$$

Here we have

$$q_U[x(F)] \leqq p_U[x(F)]\,.$$

To prove the reverse inequality we consider an equivalence class $x(F)$ with $q_U[x(F)] \leqq 1$. Then we have

$$x \in (U^0 \cap F^0)^0 = (U+F)^{00} = \overline{(U+F)}\,.$$

Consequently, for each positive number $\varepsilon$ there is an element

$$z \in \{x + \varepsilon\, U\} \cap \{U + F\}\,,$$

which we can represent in the form

$$z = x + \varepsilon x_1 \;\text{ with }\; x_1 \in U \quad\text{and}\quad z = x_2 + y \;\text{ with }\; x_2 \in U \;\text{ and }\; y \in F\,.$$

The identity

$$x - y = x_2 - \varepsilon\, x_1$$

then holds, from which the inequality

$$p_U[x(F)] \leqq p_U(x-y) = p_U(x_2 - \varepsilon\, x_1) \leqq 1 + \varepsilon$$

results. By going to the limit as $\varepsilon \to 0$, we finally obtain the assertion that for each equivalence class $x(F)$, $q_U[x(F)] \leqq 1$ always implies $p_U[x(F)] \leqq 1$. We therefore have

$$p_U[x(F)] \leqq q_U[x(F)]\,.$$

**5.1.9.** By combining the previous propositions we derive the fundamental

**Duality Theorem.** *For each nuclear $(F)$- or $(F')$-space $E$ the correspondence $F \leftrightarrow F^0$ determines a one-to-one relation between the closed linear subspaces of $E$ and $E'_b$. The following isomorphisms can be established:*

$$F = (E'_b/F^0)'_b\,, \quad E/F = (F^0)'_b\,, \quad F^0 = (E/F)'_b\,, \quad E'_b/F^0 = F'_b\,.$$

## 5.2. Topological Products and Sums

**5.2.1.** Suppose we have a set of locally convex spaces $E_i$ which correspond in a unique way to the elements of an index set $I$. The families $x = [x_i, I]$ with $x_i \in E_i$ then form a linear space $E = \prod_I E_i$ with respect to the operations

$$[x_i, I] + [y_i, I] = [x_i + y_i, I] \quad \text{and} \quad \alpha[x_i, I] = [\alpha\, x_i\, I].$$

It is easy to see that the sets

$$U = [U_i, I] = \{[x_i, I] : x_i \in U_i\} \quad \text{with} \quad U_i \in \mathfrak{U}(E_i),$$

where only finitely many zero neighborhoods $U_i$ are different from $E_i$, form a fundamental system of zero neighborhoods in $E$. The associated semi-norms are obtained from the expression

$$p_U(x) = \sup \{p_{U_i}(x_i) : i \in I\} \quad \text{with} \quad x = [x_i, I] \in E.$$

The locally convex space $E$ constructed in this way is called the **topological product** of the locally convex spaces $E_i$.

**Proposition.** *The topological product $E = \prod_I E_i$ of arbitrarily many nuclear locally convex spaces $E_i$ is also nuclear.*

*Proof.* We consider zero neighborhood $U = [U_i, I]$ of $E$ and form the finite sets

$$\mathfrak{i} = \{i \in I : U_i \neq E_i\}.$$

We now determine on the basis of Proposition 4.1.4 for $i \in \mathfrak{i}$, zero neighborhoods $V_i \in \mathfrak{U}(E_i)$ and linear forms $a_{in} \in E_i'$ with

$$\sum_I p_{V_i^0}(a_{in}) < +\infty,$$

so that the inequalities

$$p_{U_i}(x_i) \leqq \sum_I |\langle x_i, a_{in}\rangle| \quad \text{for all} \quad x_i \in E_i$$

hold. Then $V = [V_i, I]$ with $V_i = E_i$ for $i \notin \mathfrak{i}$ is a zero neighborhood of $E$.

If for each family $x = [x_i, I] \in E$ we set

$$\langle x, a_{in}\rangle = \langle x_i, a_{in}\rangle,$$

then the linear forms $a_{in}$ can be interpreted as continuous linear forms on $E$ and we have

$$\sum_{\mathfrak{i}} \sum_N p_{V^0}(a_{in}) = \sum_{\mathfrak{i}} \sum_N p_{V_i^0}(a_{in}) < +\infty.$$

Moreover, for all $x = [x_i, I] \in E$ the inequality

$$p_U(x) \leqq \sum_{\mathfrak{i}} p_{U_i}(x_i) \leqq \sum_{\mathfrak{i}} \sum_N |\langle x_i, a_{in}\rangle| = \sum_{\mathfrak{i}} \sum_N |\langle x, a_{in}\rangle|$$

holds so that the locally convex space $E$ is nuclear because of Proposition 4.1.4.

**5.2.2.** Suppose we have a set of locally convex spaces $E_i$ which correspond in a unique way to an index set $I$. The families $x = [x_i, I]$ with $x_i \in E_i$ which contain only finitely many elements $x_i \neq o$ then form a linear space $E = \sum_I E_i$ with respect to the operations

$$[x_i, I] + [y_i, I] = [x_i + y_i, I] \quad \text{and} \quad \alpha[x_i, I] = [\alpha x_i, I].$$

It is easy to see that the sets

$$U = [U_i, I] = \{[x_i, I] : x_i \in U_i\} \quad \text{with} \quad U_i \in \mathfrak{U}(E_i)$$

form a fundamental system of zero neighborhoods in $E$. The associated semi-norms are obtaind from the expression

$$p_U(x) = \sup\{p_{U_i}(x_i) : i \in I\} \quad \text{with} \quad x = [x_i, I] \in E.$$

The locally convex space $E$ constructed in this way is called the **topological direct sum** of the locally convex spaces $E_i$.

It can be shown that the sets

$$U = (U_i, I) = \left\{[\alpha_i x_i, I] : x_i \in U, \quad \sum_I |\alpha_i| \leqq 1\right\} \quad \text{with} \quad U_i \in \mathfrak{U}(E_i)$$

also form a fundamental system of zero neighborhoods in $E$. The linear space $E$ equipped with this locally convex topology is called the **locally convex direct sum** of the locally convex spaces $E_i$.

Since the relation $(U_i, I) \subset [U_i, I]$ holds, the second topology is finer than the first. In the case of a countable index set $N$ the two topologies coincide because $[2^{-n} U_n, N] \subset (U_n, N)$.

**Proposition.** *The topological (= locally convex) direct sum $E = \sum_N E_n$ of countably many nuclear locally convex spaces $E_n$ is also nuclear.*

*Proof.* We consider a zero neighborhood $U = [U_n, N]$ of $E$ and determine by Proposition 4.1.4, zero neighborhoods $V_n \in \mathfrak{U}(E_n)$ and linear forms $a_{mn} \in E_n'$ with

$$\sum_M p_{V_n^0}(a_{mn}) \leqq \frac{1}{2^n}, \qquad M = \{1, 2, \ldots\},$$

so that the inequality

$$p_{U_n}(x_n) \leqq \sum_M |\langle x_n, a_{mn}\rangle| \quad \text{for all} \quad x_n \in E_n$$

holds. Then $V = [V_n, N]$ is a zero neighborhood in $E$.

If for each family $x = [x_n, N] \in E$ we set

$$\langle x, a_{mn}\rangle = \langle x_n, a_{mn}\rangle,$$

we can then interpret the linear forms $a_{mn}$ as continuous linear forms on $E$ and we have

$$\sum_N \sum_M p_{V^\circ}(a_{mn}) = \sum_N \sum_M p_{V_n^\circ}(a_{mn}) \leqq \sum_N \frac{1}{2^n} < +\infty .$$

Since the inequality

$$p_U(x) \leqq \sum_N p_{U_n}(x_n) \leqq \sum_N \sum_M |\langle x_n, a_{mn}\rangle| \leqq \sum_N \sum_M |\langle x, a_{mn}\rangle|$$

holds for all $x = [x_n, N] \in E$, the locally convex space $E$ is nuclear by Proposition 4.1.4.

**5.2.3.** Let $E$ be a linear space which is mapped by certain linear mappings $T_i$ into locally convex spaces $E_i$ with $i \in I$ in such a way that for each element $x \neq 0$ there exists an index $i_0 \in I$ with $T_{i_0} x \neq 0$. We can then identify $E$ with that subspace of the topological product $\prod_I E_i$ which consists of all families $[T_i\, x, I]$ with $x \in E$. The locally convex topology induced on $E$ by $\prod_I E_i$ is the coarsest locally convex topology on $E$ for which all the mappings $T_i$ are continuous. The locally convex space $E$ constructed in this way is called the **locally convex kernel** of the locally convex spaces $E_i$ with respect to the mappings $T_i$.

From 5.1.1 and 5.2.1 we now obtain the

**Proposition.** *The locally convex kernel of arbitrarily many nuclear locally convex spaces is also nuclear.*

**5.2.4.** Suppose we have a set of locally convex spaces $E_i$ with $i \in I$, which are mapped into a linear space $E$ by certain linear mappings $T_i$ in such a way that $E$ is the union of the ranges of these mappings. Then $E$ can be regarded as the quotient space of the locally convex direct sum $\sum_I E_i$ by that linear subspace $F$ which is formed of all families $[x_i, I] \in \sum_I E_i$ with $\sum_I T_i\, x_i = 0$. When $F$ is closed we can equip $E$ with the associated quotient space topology which is the finest locally convex topology on $E$ for which each mapping $T_i$ is continuous. The locally convex space $E$ constructed in this way is called the **locally convex hull** of the locally convex spaces $E_i$ with respect to the mappings $T_i$.

From 5.1.3 and 5.2.2 we now obtain the

**Proposition.** *The locally convex hull of countably many nuclear locally convex spaces is also nuclear.*

## 5.3. Complete Hulls

**5.3.1.** As the most important result of this section we formulate the

**Proposition.** *The complete (quasi-complete) hull of a nuclear locally convex space E is also nuclear.*

*Proof.* It follows from the construction of the complete hull $E$, that the closures $\tilde{U}$ in $\tilde{E}$ of the zero neighborhoods $U \in \mathfrak{U}(E)$ form a fundamental system of zero neighborhoods in $\tilde{E}$.

By associating with each linear form $a \in E'$ its continuous extension $\tilde{a}$ to $\tilde{E}$, we obtain a one-to-one relation between $E'$ and $\tilde{E}'$. The two dual spaces can thus be identified with one another. For each zero neighborhood $U \in \mathfrak{U}(E)$ we have the relation $U^0 = \tilde{U}^0$, and the normed spaces $E'(\tilde{U}^0)$ and $\tilde{E}'(\tilde{U}^0)$ coincide. Consequently, the fundamental system of zero neighborhoods has property $(N)$ and the complete hull $\tilde{E}$ is nuclear.

Finally from Proposition 5.1.1 it follows that the quasi-complete hull $\hat{E}$ as a linear subspace of $\tilde{E}$ must also be nuclear.

**5.3.2.** For determining the complete hull of a nuclear metric or dual metric locally convex space we use the following

**Lemma.** *Each locally convex space $E$, in which all bounded subsets are precompact is dense in $E''_n$.*

*Proof.* For each zero neighborhood $U \in \mathfrak{U}(E)$ the strong topology on $U^0$ coincides with the weak topology because of Proposition 0.6.7. Hence $U^0$ is a compact subset of $E'_b$. Another consequence of Proposition 0.6.7 is that for each set $A \in \mathfrak{B}(E)$ the weak topology on the bipolar $A^{00}$ formed in $E''$ coincides with the natural topology.

Finally if $x$ is an arbitrary element of $E''$, there is a set $A \in \mathfrak{B}(E)$ with $x \in A^{00}$. However, because of the Bipolar Theorem, $A^{00}$ is the weakly closed hull of $A$ so that we can find in $A$ a directed system $\{x_\alpha\}$ which converges to $x$ in the weak, and thus in the natural topology.

We now obtain the

**Proposition.** *The complete (= quasi-complete) hull of a nuclear metric, resp. dual metric, locally convex space $E$ is a nuclear $(F)$-space, resp. $(F')$-space, which can be identified with $E''_b$.*

*Proof.* Since the locally convex space $E$ is quasi-barrelled by Proposition 0.7.4, resp. Lemma 4.4.12, the natural topology on $E''$ coincides with the strong topology. On the basis of the previous lemma, $E$ is a dense linear subspace of $E''_b$, so that we can regard $E''_b$ as the complete hull of $E$.

In the proof of the previous lemma we have also shown that each element $x \in E''$ is the limit of a bounded directed system of $E$. Therefore, the quasi-complete hull of $E$ coincides with complete hull.

## 5.4. Locally Concvex Tensor Produts

**5.4.1.** Although the locally convex tensor products $E \otimes_\varepsilon F$ and $E \otimes_\pi F$ of two locally convex spaces $E$ and $F$ are first described in 7.1.2, we shall prove at this point the

**Proposition.** *The locally convex tensor product $E \otimes_\varepsilon F = E \otimes_\pi F$ of two nuclear locally convex spaces $E$ and $F$ is also nuclear.*

*Proof.* Since the two locally convex tensor products $E \otimes_\varepsilon F$ and $E \otimes_\pi F$ are the same by Proposition 7.2.3, it suffices to prove our assertion for the tensor product $E \otimes_\varepsilon F$. For this purpose we consider two zero neighborhoods $U \in \mathfrak{U}(E)$ and $V \in \mathfrak{U}(F)$. Then by Proposition 4.1.4 there are zero neighborhoods $U_1 \in \mathfrak{U}(E)$ and $V_1 \in \mathfrak{U}(F)$ as well as sequences of linear forms $a_m \in E'$ and $b_n \in F'$ with

$$\sum_M p_{U_1^0}(a_m) < +\infty \quad \text{and} \quad \sum_N p_{V_1^0}(b_n) < +\infty$$

so that the estimates

$$|\langle x, a\rangle| \leqq p_U(x) \leqq \sum_M |\langle x, a_m\rangle| \quad \text{for } x \in E \text{ and } a \in U^0$$

and

$$|\langle y, b\rangle| \leqq p_V(y) \leqq \sum_N |\langle y, b_n\rangle| \quad \text{for } y \in F \text{ and } b \in V^0$$

are valid.

By means of the equation

$$\langle z, c_{mn}\rangle = \sum_{r=1}^{s} \langle x_r, a_m\rangle \langle y_r, b_n\rangle \quad \text{for } z = \sum_{r=1}^{s} x_r \otimes y_r \in E \otimes F$$

we define continuous linear forms on $E \otimes_\varepsilon F$ for which the relation

$$\sum_{M,N} \varepsilon_{(U_1,V_1)^0}(c_{mn}) \leqq \sum_{M,N} p_{U_1^0}(a_m)\, p_{V_1^0}(b_n) < +\infty \tag{1}$$

holds because of

$$|\langle z, c_{mn}\rangle| \leqq \varepsilon_{(U_1,V_1)}(z)\, p_{U_1^0}(a_m)\, p_{V_1^0}(b_n)\,.$$

Moreover, for $a \in U^0$ and $b \in V^0$, we get the estimate

$$\begin{aligned}\left|\sum_{r=1}^{s} \langle x_r, a\rangle \langle y_r, b\rangle\right| &= \left|\left\langle \sum_{r=1}^{s} x_r \langle y_r, b\rangle, a\right\rangle\right| \\ &\leqq \sum_M \left|\left\langle \sum_{r=1}^{s} x_r \langle y_r, b\rangle, a_m\right\rangle\right| = \sum_M \left|\left\langle \sum_{r=1}^{s} \langle x_r, a_m\rangle y_r, b\right\rangle\right| \\ &\leqq \sum_{M,N} \left|\left\langle \sum_{r=1}^{s} \langle x_r, a_m\rangle y_r, b_n\right\rangle\right| = \sum_{M,N} |\langle z, c_{mn}\rangle|\,,\end{aligned}$$

from which we obtain the inequality

$$\varepsilon_{(U,V)}(z) \leqq \sum_{M,N} |\langle z, c_{mn}\rangle| \quad \text{for} \quad z \in E \otimes F . \tag{2}$$

Because of Proposition 4.1.4 it follows from (1) and (2) that the locally convex space $E \otimes_\varepsilon F$ is nuclear.

**5.4.2.** As an immediate consequence of 5.3.1 and 5.4.1 we obtain the

**Proposition.** *The complete locally convex tensor product* $E \tilde{\otimes}_\varepsilon F = E \tilde{\otimes}_\pi F$ *of two nuclear locally convex spaces* $E$ *and* $F$ *is also nuclear.*

## 5.5. Spaces of Continuous Linear Mappings

**5.5.1. Proposition.** *For a dual nuclear locally convex space* $E$ *and a nuclear locally convex space* $F$ *the locally convex space* $\mathcal{L}_b(E, F)$ *is nuclear.*

*Proof.* We consider an arbitrary zero neighborhood $(A, U)$ of $\mathcal{L}_b(E, F)$ with $A \in \mathfrak{B}(E)$ and $U \in \mathfrak{U}(F)$.

Since the locally convex space $E$ is assumed to be dual nuclear there is a set $B \in \mathfrak{B}(E)$ with $A < B$ such that the canonical mapping from $E(A)$ into $E(B)$ is nuclear. Consequently there exist linear forms $\mathfrak{a}_m$ from the closed unit ball of $[E(A)]'$ and elements $x_m \in E(B)$ with

$$\sum_M p_B(x_m) < + \infty ,$$

such that for all elements $x \in E(A)$ the relation

$$x = \sum_M \langle x, \mathfrak{a}_m\rangle \, x_m$$

holds. Therefore we have,

$$|\langle x, a\rangle| \leqq \sum_M |\langle x_m, a\rangle| \quad \text{for} \quad x \in A \quad \text{and} \quad a \in E' .$$

On the basis of Proposition 4.1.4 there is a zero neighborhood $V \in \mathfrak{U}(F)$ and a sequence of continuous linear forms $b_n \in F'$ with

$$\sum_N p_{V^0}(b_n) < + \infty ,$$

such that the inequality

$$|\langle y, b\rangle| \leqq p_U(y) \leqq \sum_N |\langle y, b_n\rangle| \quad \text{for} \quad y \in F \quad \text{and} \quad b \in U^0$$

holds.

We now define a continuous linear form $A_{mn}$ on $\mathcal{L}_b(E, F)$ by

$$\langle T, A_{mn}\rangle = \langle T x_m, b_n\rangle \quad \text{for} \quad T \in \mathcal{L}(E, F) .$$

Then we obtain the estimate

$$\sum_{M,N} p_{(B,V)^0}(A_{mn}) \leqq \sum_{M,N} p_B(x_m) \, p_{V^0}(b_n) < + \infty \tag{1}$$

because

$$|\langle T, A_{mn}\rangle| \leqq p_{(B,V)}(T)\, p_B(x_m)\, p_{V^0}(b_n)\,.$$

Moreover, for $x \in A$ and $b \in U^0$ we have

$$|\langle T\,x, b\rangle| \leqq \sum_N |\langle T\,x, b_n\rangle| = \sum_N |\langle x, T'\,b_n\rangle|$$

$$\leqq \sum_{M,N} |\langle x_m, T'\,b_n\rangle| = \sum_{M,N} |\langle T, A_{mn}\rangle|\,,$$

from which the inequality

$$p_{(A,U)}(T) \leqq \sum_{M,N} |\langle T, A_{mn}\rangle| \quad \text{for} \quad T \in \mathscr{L}(E, F) \tag{2}$$

results.

From (1) and (2) it follows by Proposition 4.1.4 that the locally convex space $\mathscr{L}_b(E, F)$ is nuclear.

**5.5.2.** For two arbitrary locally convex spaces we can identify the algebraic tensor product $E'_b \otimes F$ with the linear space $\mathscr{A}(E, F)$ by identifying each element

$$z = \sum_{r=1}^{s} a_r \otimes y_r \in E'_b \otimes F$$

with the mapping

$$T\,x = \sum_{r=1}^{s} \langle x, a_r\rangle\, y_r\,.$$

Moreover, it can be shown that the $\varepsilon$-topology of $E'_b \otimes F$ coincides with the topology induced on $\mathscr{A}(E, F)$ by $\mathscr{L}_b(E, F)$.

If the locally convex space $F$ is nuclear, then it is easy to show that the finite mappings $T \in \mathscr{A}(E, F)$ are dense in $\mathscr{L}_b(E, F)$. Consequently, the locally convex space $\mathscr{L}_b(E, F)$ can be regarded as a dense linear subspace of the complete locally convex tensor product $E'_b \tilde{\otimes}_\varepsilon F$.

If we now assume that the locally convex space $E$ is dual nuclear, then $E'_b \tilde{\otimes}_\varepsilon F$ is nuclear by Proposition 5.4.2. The same assertion is then true for $\mathscr{L}_b(E, F)$ by 5.1.1. Thus we have shown that Proposition 5.5.1 can be derived from Proposition 5.4.1.

Chapter 6

# Examples of Nuclear Locally Convex Spaces

The simplest examples of nuclear locally convex spaces are the *sequence spaces* introduced by G. Köthe and O. Toeplitz which satisfy the condition presented in Theorem 6.1.2. This general criterion is due to A. Pietsch [6]. In the case of metric sequence spaces (echelon spaces), however, this had already been proven by A. Grothendieck [3].

The most important nuclear locally convex spaces consist of infinitely differentiable functions. Such spaces appeared for the first time in the *Distribution Theory* of L. Schwartz [2], where they play a very fundamental role. It was by isolating a certain property of the spaces $\mathcal{D}$ and $\mathcal{E}$ that A. Grothendieck first arrived at the concept of nuclear locally convex space.

Locally convex spaces of analytic functions have been investigated by O. Toeplitz [1], G. Köthe [3] and A. Grothendieck [7]. A. Grothendieck [3] proved such spaces are nuclear. H. G. Tillmann treated locally convex spaces consisting of harmonic functions.

We can also introduce spaces of infinitely differentiable, analytic or harmonic functions, which are defined on suitable manifolds. Moreover, in his *"théorie des courants"* De Rahm has used locally convex spaces of differential forms.

## 6.1. Sequence Spaces

**6.1.1.** We set $R = \{0, 1, 2, \ldots\}$ and consider a set $P$ of sequences of numbers $[\varrho_r, R]$ with the following properties:

($F_1$) *For all sequences* $[\varrho_r, R] \in P$ *we have* $\varrho_r \geqq 0$.

($F_2$) *For each number* $s \in R$ *there is a sequence* $[\varrho_r, R] \in P$ *with* $\varrho_s > 0$.

($F_3$) *For finitely many sequences* $[\varrho_r^{[1]}, R], \ldots, [\varrho_r^{[n]}, R] \in P$ *there exists a sequence* $[\varrho_r, R] \in P$ *with*

$$\sup \{\varrho_r^{[1]}, \ldots, \varrho_r^{[n]}\} \leqq \varrho_r \quad \textit{for all} \quad r \in R\,.$$

Under these hypotheses the collection $\Lambda = \Lambda(P)$ of all real or complex sequences of numbers $[\xi_r, R]$ with

$$\sum_R |\xi_r|\, \varrho_r < +\infty \quad \text{for} \quad [\varrho_r, R] \in P$$

forms a linear space with respect to the operations

$$[\xi_r, R] + [\eta_r, R] = [\xi_r + \eta_r, R] \quad \text{and} \quad \alpha[\xi_r, R] = [\alpha\, \xi_r, R]\,,$$

on which we can construct a locally convex topology by means of the semi-norms $p[\xi_r, R]$, given by

$$p[\xi_r, R] = \sum_R |\xi_r|\, \varrho_r \quad \text{for } [\varrho_r, R] \in P\,.$$

The locally convex space thus obtained is complete.

**6.1.2. Theorem.** *The locally convex sequence space $\Lambda(P)$ is nuclear if and only if for each sequence $[\varrho_r, R] \in P$, there are sequences $[\mu_r, R] \in \boldsymbol{l}_R^1$ and $[\sigma_r, R] \in P$ with*

$$\varrho_r \leqq \mu_r\, \sigma_r \quad \textit{for } r \in R\,.$$

*Proof.* For each zero neighborhood

$$U = \left\{[\xi_r, R] \in \Lambda : \sum_R |\xi_r|\, \varrho_r \leqq 1\right\}$$

we set

$$R(U) = \{r \in R : \varrho_r > 0\}\,.$$

We can then identify the normed space $\Lambda(U)$ with a dense linear subspace of $\boldsymbol{l}_R^1(U)$ by associating the sequence $[\xi_r\, \varrho_r, R(U)]$ with the equivalence class of the sequence $[\xi_r, R]$.

If we have two zero neighborhoods

$$U = \left\{[\xi_r, R] \in \Lambda : \sum_R |\xi_r|\, \varrho_r \leqq 1\right\}$$

and

$$V = \left\{[\xi_r, R] \in \Lambda : \sum_R |\xi_r|\, \sigma_r \leqq 1\right\}$$

with $V < U$, the canonical mapping from $\Lambda(V)$ onto $\Lambda(U)$ corresponds to the mapping from $\boldsymbol{l}_{R(V)}^1$ into $\boldsymbol{l}_{R(U)}^1$ determined by the matrix

$$[\sigma_s^{-1}\, \delta_{sr}\, \varrho_r, R(V) \times R(U)]\,.$$

By Proposition 3.1.10 the nuclearity of this mapping is equivalent to the inequality

$$\sum_{R(U)} \sigma_r^{-1}\, \varrho_r < +\infty\,.$$

This condition is satisfied precisely if the condition

$$\varrho_r \leqq \mu_r\, \sigma_r \quad \text{for all } r \in R$$

is valid for some sequence $[\mu_r, R] \in \boldsymbol{l}_R^1$.

**6.1.3.** As an easy consequence of the previous criterion we get the

**Theorem.** *A locally convex sequence space $\Lambda(P)$ is nuclear if and only if its topology can be determined by the semi-norms*

$$q[\xi_r, R] = \sup\{|\xi_r|\,\varrho_r : r \in R\} \quad \textit{with} \quad [\varrho_r, R] \in P\,.$$

**6.1.4.** From the system of all sequences of numbers $[\varrho_r, R]$ with $\varrho_r \geqq 0$ we obtain the sequence space $\Phi_R$, which consists of all sequences of numbers $[\xi_r, R]$ with only finitely many $\xi_r$ different from 0. On the other hand, from the system of sequences of numbers $[\varrho_r, R] \in \Phi_R$ with $\varrho_r \geqq 0$, we get the sequence space $\Omega_R$ which contains all sequences of numbers $[\xi_r, R]$. Both sequence spaces $\Phi_R$ and $\Omega_R$ are nuclear.

**6.1.5.** By a **power series space** we understand a sequence space $\Lambda(P)$, for which $P$ consists of all sequences of the form

$$[\varrho^{\alpha_r}, R] \quad \text{with} \quad 0 < \varrho < \varrho_0\,.$$

Here the numbers $\alpha_r$ are assumed to satisfy the condition

$$0 \leqq \alpha_0 \leqq \alpha_1 \leqq \alpha_2 \leqq \cdots.$$

We say that $\Lambda(P)$ is of **finite**, resp. **infinite**, **type** if $\varrho_0 < +\infty$, resp. $\varrho_0 = +\infty$.

By Theorem 6.1.2 in order for such a spaces to be nuclear when $\varrho_0 < +\infty$, resp. $\varrho_0 = +\infty$, it is necessary and sufficient that the inequality

$$\sum_R q^{\alpha_r} < +\infty$$

hold for each, resp. some, number $q$ between 0 and 1.

The simplest power series of finite or infinite type are obtained from the numbers $\alpha_r = r$.

**6.1.6.** We get the most important nuclear power series space $\Sigma$ from the system of sequences

$$\{[\varrho^{\log(r+1)}, R] : 0 < \varrho < +\infty\} \quad \text{or} \quad \{[(r+1)^n, R] : n \in N\}\,.$$

$\Sigma$ then consists precisely of all sequences of numbers $[\xi_r, R]$ with

$$\lim_r\,(r+1)^n\,|\xi_r| = 0 \quad \text{for} \quad n \in N\,.$$

## 6.2. Spaces of Infinitely Differentiable Functions

**6.2.1.** In the sequel we denote by $\mathscr{E}(\Delta)$ the collection of all infinitely differentiable real or complex functions $f = [f(t)]$ on the closed interval $\Delta = [a, b]$, which form a linear space with respect to the operations

$$[f(t)] + [g(t)] = [f(t) + g(t)] \quad \text{and} \quad \alpha[f(t)] = [\alpha\, f(t)]\,.$$

It is easily seen that $\mathscr{E}(\Delta)$ is an $(F)$-space when we use the locally convex topology determined by the norms

$$p_r(f) = \sup\left\{\sum_{k=0}^{r} |f^{[k]}(t)| : a \leqq t \leqq b\right\} \quad \text{with} \quad r = 0, 1, \ldots .$$

**Theorem.** *The locally convex space $\mathscr{E}(\Delta)$ is nuclear.*

*Proof.* For each function $f \in \mathscr{E}(\Delta)$ we have the identity

$$f^{[k]}(t) - f^{[k]}(a) = \int_a^t f^{[k+1]}(s)\, ds$$

from which we obtain the estimate

$$|f^{[k]}(t)| \leqq \int_a^b |f^{[k+1]}(s)|\, ds + |f^{[k]}(a)| .$$

However, since the right side of the last relation no longer depends on $t$, we get the inequality

$$p_r(f) \leqq \sum_{k=0}^{r} \left\{\int_a^b |f^{[k+1]}(s)|\, ds + |f^{[k]}(a)|\right\} .$$

by summing over $k = 0, 1, \ldots, r$ and then forming the supremum.

Since the continuous linear forms $\delta_s^{[k]}$ with

$$\langle f, \delta_s^{[k]}\rangle = f^{[k]}(s)$$

for $s \in \Delta$ and $k = 0, 1, \ldots, r+1$ are in the polar set of the zero neighborhood

$$V = \{f \in \mathscr{E}(\Delta) : p_{r+1}(f) \leqq 1\}$$

it is possible to define a positive Radon measure on $V^0$ by

$$\int_{V^0} \varphi(a)\, d\mu = \sum_{k=0}^{r} \left\{\int_a^b \varphi(\delta_s^{[k+1]})\, ds + \varphi(\delta_a^{[k]})\right\} \quad \text{for} \quad \varphi \in \mathscr{C}(V^0)$$

so that the estimate

$$p_r(f) \leqq \int_{V^0} |\langle f, a\rangle| d\mu \quad \text{for all} \quad f \in \mathscr{E}(\Delta)$$

is valid. Consequently, the locally convex space $\mathscr{E}(\Delta)$ is nuclear by Proposition 4.1.5.

**6.2.2.** The linear space $\mathscr{E}$ of all infinitely differentiable real or complex functions $f = [f(t)]$ on the line $(-\infty, +\infty)$ is an $(F)$-space if we equip it with the locally convex topology obtained from the semi-norms

$$q_r(f) = \sup\left\{\sum_{k=0}^{r} |f^{[k]}(t)| : -r \leqq t \leqq +r\right\} \quad \text{with} \quad r = 0, 1, \ldots .$$

If the restriction of each function $f \in \mathscr{E}$ to the interval $\Delta_r = [-r, +r]$ is denoted by $T_r[f]$ then it can be shown that $\mathscr{E}$ is the locally convex kernel of the nuclear $(F)$-spaces $\mathscr{E}(\Delta_r)$ with respect to the mappings $T_r$ with $r = 0, 1, \ldots$. Therefore, by Proposition 5.2.3 we have the

**Theorem.** *The locally convex space $\mathscr{E}$ is nuclear.*

**6.2.3.** If we denote by $\mathscr{E}_0(\Delta)$ that linear subspace of $\mathscr{E}(\Delta)$, which consists of all functions $f$ with $f^{(k)}(a) = f^{(k)}(b) = 0$ for $k = 0, 1, \ldots$, then by Proposition 5.1.1 we get the

**Theorem.** *The locally convex space $\mathscr{E}_0(\Delta)$ is nuclear.*

**6.2.4.** We now consider a space basic to distribution theory, the linear space $\mathscr{D}$ of all infinitely differentiable real or complex functions $f = [f(t)]$ on the line $(-\infty, +\infty)$ which vanish outside of an interval $\Delta_r = [-r, +r]$ (depending on $f$).

By associating with each function $f \in \mathscr{E}_0(\Delta_r)$ the function $T_r[f]$, which is equal to $f(t)$ on $\Delta_r$ and equal to 0 outside of $\Delta_r$, we obtain $\mathscr{D}$ as the union of the ranges of the mappings $T_r$. It can be shown that a locally convex topology can be introduced on the linear space $\mathscr{D}$ in the sense of 5.2.4. The locally convex space thus obtained is complete but not metric or dual metric. Moreover by Proposition 5.2.4 we have the

**Theorem.** *The locally convex space $\mathscr{D}$ is nuclear.*

**6.2.5.** The linear space $\mathscr{S}$ of all infinitely differentiable real or complex functions $f = [f(t)]$ on the line $(-\infty, +\infty)$ with the norms

$$s_r(f) = \sup\left\{(1+t^2)^r \sum_{k=0}^{r} |f^{[k]}(t)| : -\infty < t < +\infty\right\} < +\infty \quad \text{for } r = 0, 1, \ldots$$

is an $(F)$-space if we give it the topology obtained from the norms $s_r(f)$.

**Theorem.** *The locally convex space $\mathscr{S}$ is nuclear.*

*Proof.* For each function $f \in \mathscr{S}$ the identity

$$(1+t^2)^r f^{[k]}(t) = \int_{-\infty}^{t} \{2\, r\, s\, (1+s^2)^{r-1} f^{[k]}(s) + (1+s^2)^r f^{[k+1]}(s)\}\, ds$$

leads to the estimate

$$(1+t^2)^r |f^{[k]}(t)| \leq \int_{-\infty}^{+\infty} \{r\, (1+s^2)^r |f^{[k]}(s)| + (1+s^2)^r |f^{[k+1]}(s)\}\, ds$$

since $|2\, s| \leq (1+s^2)$. The right side of this relation no longer depends on $t$ so that by summing over $k = 0, 1, \ldots, r$ and then forming the supremum we get the inequality

$$s_r(f) \leq \sum_{k=0}^{r} \int_{-\infty}^{+\infty} r\, (1+s^2)^r |f^{[k]}(s)| + (1+s^2)^r |f^{[k+1]}(s)|\}\, ds\,.$$

We consequently have

$$s_r(f) \leqq (r+1) \sum_{k=0}^{r+1} \int_{-\infty}^{+\infty} (1+s^2)^r \, |f^{[k]}(s)| \, ds \, .$$

Since the continuous linear forms $\varepsilon_s^{[k]}$ with

$$\langle f, \varepsilon_s^{[k]} \rangle = (1+s^2)^{r+1} f^{[k]}(s)$$

for $-\infty < s < +\infty$ and $k = 0, 1, \ldots, r+1$ lie in the polar set of the zero neighborhood,

$$V = \{f \in \mathscr{S} : s_{r+1}(f) \leqq 1\}$$

it is possible to define a positive Radon measure on $V^0$ by the equation

$$\int_{V^0} \varphi(a) \, d\mu = (r+1) \sum_{k=0}^{r+1} \int_{-\infty}^{+\infty} \varphi(\varepsilon_s^{[k]}) \, (1+s^2)^{-1} \, ds \qquad \text{for } \varphi \in \mathscr{C}(V^0)$$

such that the estimate

$$s_r(f) \leqq \int_{V^0} |\langle f, a \rangle| \, d\mu \qquad \text{for all } f \in \mathscr{S}$$

holds. Consequently, the locally convex space $\mathscr{S}$ is nuclear by Proposition 4.1.5.

**6.2.6.** Finally we obtain the

**Theorem.** *The strong topological duals of the locally convex spaces* $\mathscr{E}(\Delta)$, $\mathscr{E}_0(\Delta)$, $\mathscr{E}$, $\mathscr{S}$ *and* $\mathscr{D}$ *are nuclear.*

*Proof.* Since all of the locally convex spaces introduced except $\mathscr{D}$ are metric, our conclusion essentially results from Theorem 4.3.3. Since it can be shown that the locally convex space $\mathscr{D}$ has property $(B)$, the nuclearity of $\mathscr{D}'$ follows from Theorem 4.3.1. But it can also be seen that $\mathscr{D}'$ is the locally convex kernel of the nuclear $(F')$-spaces $\mathscr{E}_0(\Delta_r)'$ with respect to the mappings $T_r'$.

**6.2.7.** The *method of reproducing kernels* developed by J. Włoka [1], [2] can be used to demonstrate the nuclearity of a large class of function and distribution spaces.

## 6.3. Spaces of Harmonic Functions

**6.3.1.** In the sequel we denote a point of $n$-dimensional Euclidean space $R^n$ by $t = [t_1, \ldots, t_n]$ and the distance between the points $s$ and $t$ by

$$\varrho(s, t) = \left\{ \sum_{k=1}^{n} |s_k - t_k|^2 \right\}^{1/2} .$$

The mean value of a continuous real or complex function $f = [f(t)]$ over the ball

$$K(t, \varepsilon) = \{s \in R^n : \varrho(s, t) \leqq \varepsilon\}$$

is the integral

$$M(f, t, \varepsilon) = \frac{1}{\varepsilon^n V_n} \int_{K(t,\varepsilon)} f(s)\, ds .$$

Here $V_n$ is the volume of the $n$-dimensional unit ball.

**6.3.2.** A continuous real or complex function $f = [f(t)]$ is called **harmonic** on the open subset $G$ of $R^n$ if for each ball $K(t, \varepsilon) \subset G$ the identity

$$f(t) = M(f, t, \varepsilon)$$

is valid.

The collection $\mathscr{H}(G)$ of all harmonic functions on $G$ is a linear space with respect to the operations

$$[f(t)] + [g(t)] = [f(t) + g(t)] \quad \text{and} \quad \alpha[f(t)] = [\alpha\, f(t)]$$

on which we can construct a locally convex topology from the seminorms

$$p_K(f) = \sup\{|f(t)| : t \in K\} ,$$

in which $K$ ranges over all compact subsets of $G$.

We now consider the special compact subsets $K_r$ with $r = 1, 2, \ldots$ which consist of all points $t \in G$ with

$$\varrho(t, 0) \leqq r \quad \text{and} \quad \varrho(s, t) \geqq 1/r \quad \text{for} \quad s \notin G .$$

Since every arbitrary compact subset $K$ of $G$ lies in almost every set $K_r$, we see that the topology of $\mathscr{H}(G)$ can also be constructed from the monotonically increasing sequence of seminorms

$$p_r(f) = \sup\{|f(t)| : t \in K_r\} .$$

As a simple consequence of the mean value property of harmonic functions it follows that $\mathscr{H}(G)$ is complete and hence an $(F)$-space.

**6.3.3. Theorem.** *The locally convex space $\mathscr{H}(G)$ is nuclear.*

*Proof.* For $K$ an arbitrary compact subset of $G$ we determine a positive number $\varepsilon$ such that the set

$$H = \bigcup\{K(t, \varepsilon) : t \in K\}$$

which is likewise compact, also lies in $G$. Then because of the mean value property the estimate

$$|f(t)| \leqq \frac{1}{\varepsilon^n V_n} \int_{K(t,\varepsilon)} |f(s)|\, ds \leqq \frac{1}{\varepsilon^n V_n} \int_H |f(s)|\, ds ,$$

is true for $t \in K$ and all harmonic functions $f \in \mathscr{H}(G)$. From this estimate we obtain the inequality

$$p_K(f) \leqq \frac{1}{\varepsilon^n V_n} \int_H |f(s)| \, ds .$$

Since the continuous linear forms $\delta_s$ with

$$\langle f, \delta_s \rangle = f(s)$$

for $s \in H$ lie in the polar set of the zero neighborhood

$$V = \{f \in \mathscr{H}(G) : p_H(f) \leqq 1\}$$

it is possible to define a positive Radon measure $\mu$ on $V^0$ by

$$\int_{V^0} \varphi(a) \, d\mu = \frac{1}{\varepsilon^n V_n} \int_H \varphi(\delta_s) \, ds \quad \text{for } \varphi \in \mathscr{C}(V^0)$$

for which the estimate

$$p_K(f) \leqq \int_{V^0} |\langle f, a \rangle| \, d\mu \quad \text{for all } f \in \mathscr{H}(G)$$

holds. Consequently, the locally convex space $\mathscr{H}(G)$ is nuclear by Proposition 4.1.5.

**6.3.4.** The nuclearity of the locally convex spaces encountered in axiomatic potential theory was proven independently by P. A. Loeb and B. Walsh [1], resp. H. Bauer [1].

**6.3.5.** It was shown by Y. Kōmura [1] that the solution space of a homogeneous partial differential equation with constant coefficients is nuclear precisely in the hypoelliptic case. Related results are found in J. Włoka [3] and A. Pietsch [11].

**6.3.6.** We now formulate a general nuclearity criterion for spaces of continuous functions.

If $\mathscr{C}(G)$ is the $(F)$-space of all continuous real or complex functions on $G$ with the topology constructed from the semi-norms,

$$p_K(f) = \sup \{|f(t)| : t \in K\} ,$$

then according to A. Pietsch [11], [13] we have the

**Theorem.** *A linear subspace $\mathscr{N}(G)$ of $\mathscr{C}(G)$ is nuclear if and only if there is a positive Radon measure $\mu$ on $G$ such that the topology of $\mathscr{N}(G)$ is determined by the semi-norms*

$$s_K(f) = \left\{\int_K |f(t)|^p \, d\mu\right\}^{1/p} \quad (1 \leqq p < +\infty, \ p \text{ fixed}) .$$

## 6.4. Spaces of Analytic Functions

**6.4.1.** The collection $\mathcal{A}(G)$ of all analytic functions $f = [f(z)]$ on an open set of the finite complex plane is a linear space with respect to the operations

$$[f(z)] + [g(z)] = [f(z) + g(z)] \quad \text{and} \quad \alpha[f(z)] = [\alpha\, f(z)] .$$

We can construct on $\mathcal{A}(G)$ a locally convex topology from the semi-norms

$$p_K(f) = \sup \{|f(z)| : z \in K\}$$

in which $K$ ranges over all compact subsets of $G$.

We now consider the special compact subsets $K_r$ with $r = 1, 2, \ldots$ which consist of all complex numbers $z \in G$ with

$$|z| \leqq r \quad \text{and} \quad |z - w| \geqq \frac{1}{r} \quad \text{for } w \notin G .$$

Since each arbitrary compact subset $K$ of $G$ lies in almost every set $K_r$ we see that the topology of $\mathcal{A}(G)$ can also be constructed from the monotonically increasing sequence of semi-norms

$$p_r(f) = \sup \{|f(z)| : z \in K_r\} .$$

It follows from the Weierstrass convergence theorem that $\mathcal{A}(G)$ is complete and hence an $(F)$-space.

**6.4.2. Theorem.** *The locally convex space $\mathcal{A}(G)$ is nuclear.*

*Proof.* We first consider an arbitrary disc

$$K = \{w : |w - z_0| \leqq \varrho\} ,$$

contained in $G$ and determine a positive number $\varepsilon$, such that the disc

$$H = \{w : |w - z_0| \leqq \varrho + \varepsilon\}$$

is also contained in $G$. Then, on the basis of the Cauchy integral formula

$$f(z) = \frac{1}{2\pi i} \int\limits_{|w - z_0| = \varrho + \varepsilon} \frac{f(w)}{w - z}\, dw$$

we have the estimate

$$|f(z)| \leqq \frac{\varrho + \varepsilon}{2\pi\varepsilon} \int\limits_0^{2\pi} |f(w)|\, d\alpha \quad \text{with } w = z_0 + (\varrho + \varepsilon)\, e^{i\alpha} .$$

for each complex $z \in K$ and all analytic functions $f \in \mathcal{A}(G)$ because $|w - z| \leqq \varepsilon$. Consequently, we also have

$$p_K(f) \leqq \frac{\varrho + \varepsilon}{2\pi\varepsilon} \int\limits_0^{2\pi} |f(w)|\, d\alpha .$$

Since the continuous linear forms $\delta_w$ with

$$\langle f, \delta_w \rangle = f(w) \quad \text{for} \quad |w - z_0| = \varrho + \varepsilon$$

lie in the polar set of the zero neighborhood

$$V = \{f \in \mathcal{A}(G) : p_H(f) \leqq 1\},$$

it is possible to define a positive Radon measure $\mu$ on $V^0$ by

$$\int_{V^0} \varphi(a)\, d\mu = \frac{\varrho + \varepsilon}{2\pi\varepsilon} \int_0^{2\pi} \varphi(\delta_w)\, d\alpha \qquad \text{for} \quad \mathscr{C}(V^0)$$

for which the inequality

$$p_K(f) \leqq \int_{V^0} |\langle f, a \rangle|\, d\mu \qquad \text{for all} \quad f \in \mathcal{A}(G)$$

is valid.

Now if $K$ is an arbitrary compact subset of $G$, we cover it by finitely many discs $K_i$. By the argument above there are zero neighborhoods $V_i$ and positive Radon measures $\mu_i$ with

$$p_{K_i}(f) \leqq \int_{V_i^0} |\langle f, a \rangle|\, d\mu_i \qquad \text{for all} \quad f \in \mathcal{A}(G),$$

and we can define a positive Radon measure $\mu$ on the polar of the zero neighborhood $V = \bigcap V_i$ by the equation

$$\int_{V^0} \varphi(a)\, d\mu = \sum \int_{V_i^0} \varphi(a)\, d\mu_i \qquad \text{for all} \quad \varphi \in \mathscr{C}(V^0)$$

such that for all analytic functions $f \in \mathcal{A}(G)$ the estimate

$$p_K(f) \leqq \sum p_{K_i}(f) \leqq \sum \int_{V_i^0} |\langle f, a \rangle|\, d\mu_i = \int_{V^0} |\langle f, a \rangle|\, d\mu$$

is valid. Consequently the locally convex space $\mathcal{A}(G)$ is nuclear by 4.1.5.

Chapter 7

# Locally Convex Tensor Products

The concept of topological tensor product of two Hilbert spaces appeared for the first time in the works of F. J. Murray and J. von Neumann, 1936/37. The extension of this idea to Banach spaces was undertaken by R. Schatten [1]. A detailed collection of all further investigations is found in a monograph of R. Schatten [3]. The general *locally convex tensor product* was introduced by A. Grothendieck [1], [2], [3]. It forms the essential basis for his theory of nuclear locally convex spaces. In contrast to this, locally convex tensor products are not used for our development of this theory. Nevertheless, they will be briefly treated in this chapter.

We begin by defining the algebraic tensor product of two locally convex spaces on which we then introduce two locally convex topologies which in general are distinct. In section 7.2 we relate the results of 4.2 to tensor products by showing that for each complete locally convex space $E$ the locally convex tensor product $\boldsymbol{l}_I^1 \tilde{\otimes}_\pi E$, resp. $\boldsymbol{l}_I^1 \tilde{\otimes}_\varepsilon E$, can be identified with the space $\boldsymbol{l}_I^1\{E\}$, resp. $\boldsymbol{l}_I^1(E)$. These statements were first proved by A. Grothendieck [3] as well as the characterization of nuclear locally convex spaces presented in Theorem 7.3.1.

L. Schwartz was able to show in 1948 that for each continuous linear mapping $T$ from $\mathfrak{D}$ into $\mathfrak{D}'$ there is a uniquely determined distribution $[T(x, y)]$ such that

$$T(\varphi) = [\int T(x, y)\, \varphi(y)\, dy] \quad \text{for} \quad \varphi \in \mathfrak{D}\,.$$

All such mappings can thus be represented in terms of a "*kernel*". The question of the general validity of such representations led A. Grothendieck to the concept of nuclear convex spaces.

In the sequel we omit the classical "*théoremè des noyaux*" and refer the reader to the book of I. M. Gelfand and N. J. Vilenkin as well as the works of W. Bogdanowicz, K. Ehrenpreis and H. Gask. Thus in 7.4.3 we present an abstract version of the Kernel Theorem which leads to a further characterization of nuclear locally convex spaces (Theorem 7.4.4.)

The last section of this chapter is presented only for its historical content. It shows how we can fit the concept of nuclear mapping into the framework of locally convex tensor products. We should, perhaps, call attention to a very simple new proof of Theorem 7.5.1.

## 7.1. Definition of Locally Convex Tensor Products

**7.1.1.** The algebraic tensor product $E \otimes F$ of two linear spaces $E$ and $F$ is the linear space of all formal finite sums

$$z = \sum_{r=1}^{n} x_r \otimes y_r \quad \text{with} \quad x_r \in E \quad \text{and} \quad y_r \in F\,,$$

where the following expressions are identified:

$$(x_1 + x_2) \otimes y = x_1 \otimes y + x_2 \otimes y\,,\ x \otimes (y_1 + y_2) = x \otimes y_1 + x \otimes y_2\,,$$
$$\alpha\,(x \otimes y) = (\alpha\, x) \otimes y = x \otimes (\alpha\, y)\,.$$

**7.1.2.** Let $E$ and $F$ be two locally convex spaces. For arbitrary zero neighborhoods $U \in \mathfrak{U}(E)$ and $V \in \mathfrak{U}(F)$ we set

$$\pi_{(U,V)}(z) = \inf\left\{\sum_{r=1}^{n} p_U(x_r)\, p_V(y_r)\right\}.$$

Here the infimum is taken over all possible representations of the element $z$ in the form

$$z = \sum_{r=1}^{n} x_r \otimes y_r \quad \text{with} \quad x_r \in E \quad \text{and} \quad y_r \in F\,.$$

On the other hand, the expression

$$\varepsilon_{(U,V)}(z) = \sup\left\{\left|\sum_{r=1}^{n} \langle x_r, a\rangle \langle y_r, b\rangle\right| : a \in U^0,\ b \in V^0\right\}$$

is independent of the special representation of the element $z$.

By means of these equations we obtain on the algebraic tensor product $E \otimes F$, two systems of semi-norms from which locally convex topologies ($\pi$-topology and $\varepsilon$-topology) can be constructed. The locally convex tensor products obtained in this way are denoted by $E \otimes_\pi F$ and $E \otimes_\varepsilon F$.

**7.1.3. Proposition.** *On the algebraic tensor product $E \otimes F$ of two locally convex spaces $E$ and $F$ the $\pi$-topology is finer than the $\varepsilon$-topology.*

*Proof.* Let $U \in \mathfrak{U}(E)$ and $V \in \mathfrak{U}(F)$ be two arbitrary zero neighborhoods then the inequality

$$\left|\sum_{r=1}^{n} \langle x_r, a\rangle \langle y_r, b\rangle\right| \leq \sum_{r=1}^{n} p_U(x_r)\, p_V(y_r)\,.$$

holds for each element

$$z = \sum_{r=1}^{n} x_r \otimes y_r \in E \otimes F$$

and all linear forms $a \in U^0$, $b \in V^0$. Thus we have

$$\boldsymbol{\varepsilon}_{(U,V)}(z) \leqq \sum_{r=1}^{n} p_U(x_r)\, p_V(y_r) \,.$$

But since this estimate holds for each representation of the element $z$, we obtain the statement

$$\boldsymbol{\varepsilon}_{(U,V)}(z) \leqq \boldsymbol{\pi}_{(U,V)}(z) \qquad \text{for all} \quad z \in E \otimes F \,.$$

## 7.2. Special Locally Convex Tensor Products

**7.2.1.** In this section we investigate the locally convex tensor product of the Banach space $\boldsymbol{l}_I^1$ with an arbitrary locally convex space $E$. As the first result we have the

**Proposition.** *The algebraic tensor product $\boldsymbol{l}_I^1 \otimes E$ can be identified with the linear space of all finite dimensional summable (= absolutely summable) families $[x_i, I]$ from the locally convex space $E$.*

*Proof.* We first show that the correspondence

$$\sum_{r=1}^{n} [\xi_i^{[r]}, I] \otimes y_r \to \left[ \sum_{r=1}^{n} \xi_i^{[r]} y_r, I \right]$$

is a one-to-one mapping $H$ from $\boldsymbol{l}_I^1 \otimes E$ into the linear space of all families $[x_i, I]$ from $E$. For this purpose we consider an element

$$\sum_{r=1}^{n} [\xi_i^{[r]}, I] \otimes y_r \in \boldsymbol{l}_I^1 \otimes E \,,$$

which is mapped into the family $[o, I]$ by $H$. We then have

$$\sum_{r=1}^{n} \xi_i^{[r]} y_r = o \qquad \text{for} \quad i \in I \,.$$

We now determine finitely many linearly independent elements $e_1, \ldots, e_m$, of which the elements $y_1, \ldots, y_n$ are linear combinations:

$$y_r = \sum_{s=1}^{m} \alpha_{rs}\, e_s \qquad (r = 1, \ldots, n) \,.$$

If we set

$$\eta_i^{[s]} = \sum_{r=1}^{n} \xi_i^{[r]} \alpha_{rs} \qquad (s = 1, \ldots, m \quad \text{and} \quad i \in I) \,,$$

then

$$\sum_{s=1}^{m} \eta_i^{[s]} e_s = o \qquad \text{for all} \quad i \in I .$$

Consequently, we have

$$\eta_i^{[s]} = 0 \qquad \text{for} \quad s = 1, \ldots, m \quad \text{and} \quad i \in I .$$

We now obtain

$$\sum_{r=1}^{n} [\xi_i^{[r]}, I] \otimes y_r = \sum_{s=1}^{m} [\eta_i^{[s]}, I] \otimes e_s = [o, I] \otimes o .$$

We have thus shown that the linear mapping $H$ is one-to-one.

It is clear that every family from the range of $H$ is finite dimensional and absolutely summing. On the other hand, each finite dimensional summable family $[x_i, I]$ can be written in the form

$$x_i = \sum_{r=1}^{n} \xi_i^{[r]} e_r \qquad \text{with} \quad [\xi_i^{[r]}, I] \in \boldsymbol{l}_I^1 \quad \text{and} \quad e_r \in E$$

because of the considerations made in 1.6.2. Consequently, we have

$$H\left( \sum_{r=1}^{n} [\xi_i^{[r]}, I] \otimes e_r \right) = [x_i, I] .$$

**7.2.2. Proposition.** *For each locally convex space $E$ the locally convex tensor product $\boldsymbol{l}_I^1 \otimes_\pi E$ can be identified with that linear subspace of $\boldsymbol{l}_I^1\{E\}$ which consists of finite dimensional families.*

*Proof.* Since the algebraic part of our assertion follows immediately from 7.2.1, we have only to show that the $\boldsymbol{\pi}$-topology of $\boldsymbol{l}_I^1 \otimes E$ coincides with the $\boldsymbol{\pi}$-topology induced by $\boldsymbol{l}_I^1\{E\}$. For this we prove the identity

$$\boldsymbol{\pi}_U[x_i, I] = \inf \left\{ \sum_{r=1}^{n} \boldsymbol{\lambda}_1[\xi_r^{[r]}, I]\, p_U(y_r) \right\}$$

for each finite dimensional family $[x_i, I] \in \boldsymbol{l}_I^1\{E\}$ and all zero neighborhoods $U \in \mathfrak{U}(E)$. Here the infimum on the right-hand side, which we denote by $\boldsymbol{\pi}_{(\lambda_1, U)}[x_i, I]$ is taken over all possible representations of the family $[x_i, I]$ in the form

$$x_i = \sum_{r=1}^{n} \xi_i^{[r]} y_r \qquad \text{with} \quad [\xi_i^{[r]}, I] \in \boldsymbol{l}_I^1 \quad \text{and} \quad y_r \in E .$$

Since the inequality

$$\boldsymbol{\pi}_U[x_i, I] = \sum_I p_U(x_i) \leqq \sum_I \sum_{r=1}^{n} |\xi_i^{[r]}|\, p_U(y_r) = \sum_{r=1}^{n} \boldsymbol{\lambda}_1 [\xi_i^{[r]}, I]\, p_U(y_r)$$

holds, we have

$$\boldsymbol{\pi}_U[x_i, I] \leqq \boldsymbol{\pi}_{(\lambda_1, U)}[x_i, I] .$$

We now consider any representation of the family $[x_i, I]$ in the form

$$x_i = \sum_{r=1}^{n} \xi_i^{[r]} y_r \quad \text{with} \quad [\xi_i^{[r]}, I] \in \boldsymbol{l}_I^1 \quad \text{and} \quad y_r \in E$$

and determine for each positive number $\delta$ a set $\mathfrak{i} \in \mathfrak{F}(I)$ for which the estimates

$$\sum_{I \setminus \mathfrak{i}} |\xi_i^{[r]}| \, p_U(y_r) \leq \frac{\delta}{n} \quad \text{for } r = 1, \ldots, n$$

are valid.

We then have

$$\boldsymbol{\pi}_{(\lambda_1, U)}[x_i - x_i(\mathfrak{i}), I] \leq \sum_{r=1}^{n} \boldsymbol{\lambda}_1 [\xi_i^{[r]} - \xi_i^{[r]}(\mathfrak{i}), I] \, p_U(y_r) \leq \delta$$

and

$$\boldsymbol{\pi}_{(\lambda_1, U)}[x_i(\mathfrak{i}), I] \leq \sum_{\mathfrak{i}} p_U(x_i) \leq \boldsymbol{\pi}[x_i, I] .$$

Consequently, the relation

$$\boldsymbol{\pi}_{(\lambda_1, U)}[x_i, I] \leq \boldsymbol{\pi}_{(\lambda_1, U)}[x_i - x_i(\mathfrak{i}), I] + \boldsymbol{\pi}_{(\lambda_1, U)}[x_i(\mathfrak{i}), I] \leq \delta + \boldsymbol{\pi}_U[x_i, I] ,$$

holds, and from it we obtain the inequality

$$\boldsymbol{\pi}_{(\lambda_1, U)}[x_i, I] \leq \boldsymbol{\pi}_U[x_i, I]$$

by taking the limit as $\delta \to 0$.

Thus for all finite dimensional families $[x_i, I] \in \boldsymbol{l}_I^1\{E\}$ the identity

$$\boldsymbol{\pi}_U[x_i, I] = \boldsymbol{\pi}_{(\lambda_1, U)}[x_i, I]$$

is proven.

**7.2.3. Theorem.** *For each complete locally convex space $E$ the complete locally convex tensor product $\boldsymbol{l}_I^1 \tilde{\otimes}_\pi E$ can be identified with the locally convex space $\boldsymbol{l}_I^1\{E\}$.*

*Proof.* If we consider the locally convex tensor product $\boldsymbol{l}_I^1 \otimes_\pi E$ as a linear subspace of $\boldsymbol{l}_I^1\{E\}$, we obtain from the relation

$$\boldsymbol{\pi}\text{-}\lim_{\mathfrak{i}} [x_i(\mathfrak{i}), I] = [x_i, I] ,$$

that the linear space $\boldsymbol{l}_I^1 \otimes E$ is dense in $\boldsymbol{l}_I^1\{E\}$. But since the locally convex space $\boldsymbol{l}_I^1\{E\}$ is complete by Proposition 1.4.3, we can identify it with the complete hull of $\boldsymbol{l}_I^1 \otimes_\pi E$.

**7.2.4. Proposition.** *For each locally convex space $E$ the locally convex tensor product $\boldsymbol{l}_I^1 \otimes_\varepsilon E$ can be identified with that linear subspace of $\boldsymbol{l}_I^1(E)$ which consists of the finite dimensional families.*

*Proof.* Since the algebraic part of our assertion follows immediately from 7.2.1, we have only to show that the $\boldsymbol{\varepsilon}$-topology of $\boldsymbol{l}_I^1 \otimes E$ coincides with the $\boldsymbol{\varepsilon}$-topology induced by $\boldsymbol{l}_I^1(E)$.

For this purpose we consider an arbitrary finite dimensional summable family $[x_i, I]$ from $E$ and represent it in the form

$$x_i = \sum_{r=1}^{n} \xi_i^{[r]} y_r \quad \text{with} \quad [\xi_i^{[r]}, I] \in \mathfrak{l}_I^1 \quad \text{and} \quad y_i \in E .$$

If $U$ is a zero neighborhood from $\mathfrak{U}(E)$ we have

$$\varepsilon_U[x_i, I] = \sup \left\{ \sum_I |\langle x_i, a \rangle| : a \in U^0 \right\}$$

by definition. Consequently, the relation

$$\varepsilon_U[x_i, I] = \sup \left\{ \left| \sum_I \alpha_i \langle x_i, a \rangle \right| : |\alpha_i| \leqq 1, \ a \in U^0 \right\}$$

holds. Now we obtain the identity

$$\varepsilon_U[x_i, I] = \sup \left\{ \left| \sum_{r=1}^{n} \langle [\xi_i^{[r]}, I], [\alpha_i, I] \rangle \langle y_r, a \rangle \right| : |\alpha_i| \leqq 1, \ a \in U^0 \right\}$$

by the substitution

$$x_i = \sum_{r=1}^{n} \xi_i^{[r]} y_r .$$

A comparison with the definition given in 7.1.2 shows that the right side of our equation coincides with the $\varepsilon_{(\lambda_1, U)}$-semi-norm of the family $[x_i, I]$ formed in $\mathfrak{l}_I^1 \otimes E$. Thus our assertion is proven.

**7.2.5.** From the same considerations as in 7.2.3 we now get the

**Theorem.** *For each complete locally convex space E the complete locally convex tensor product* $\mathfrak{l}_I^1 \tilde{\otimes} E$ *can be identified with the locally convex space* $\mathfrak{l}_I^1(E)$.

**7.2.6.** We now prove the

**Proposition.** *For each locally convex space E the statements*

$$\mathfrak{l}_I^1 \otimes_\varepsilon E \equiv \mathfrak{l}_I^1 \otimes_\pi E \quad (1) \qquad \textit{and} \qquad \mathfrak{l}_I^1(E) \equiv \mathfrak{l}_I^1\{E\} \qquad (2)$$

*are equivalent.*

*Proof.* Since the implication (2) → (1) is trivial, we have only to show that (1) also follows from (2).

For this purpose we assume that for a locally convex space $E$ the assertion (1) is valid. Then the $\varepsilon$-topology on $\mathfrak{l}_I^1 \otimes E$ coincides with the $\pi$-topology and for each zero neighborhood $U \in \mathfrak{U}(E)$ there is a zero neighborhood $V \in \mathfrak{U}(E)$ with

$$\pi_U[x_i, I] \leqq \varepsilon_V[x_i, I] \quad \text{for} \quad [x_i, I] \in \mathfrak{l}_I^1 \otimes E .$$

Since for each summable family $[x_i, I]$ from $E$ the families $[x_i(\mathfrak{i}), I]$ with $\mathfrak{i} \in \mathfrak{F}(I)$ belong to $\boldsymbol{l}_I^1 \otimes E$, we have

$$\sum_{\mathfrak{i}} p_U(x_i) = \boldsymbol{\pi}_U[x_i(\mathfrak{i}), I] \leqq \boldsymbol{\varepsilon}_V[x_i(\mathfrak{i}), I] \leqq \boldsymbol{\varepsilon}_V[x_i, I] .$$

Consequently, the inequality

$$\boldsymbol{\pi}_U[x_i, I] = \sum_I p_U(x_i) \leqq \boldsymbol{\varepsilon}_V[x_i, I] ,$$

holds, and the family $[x_i, I]$ is also absolutely summable. Therefore, the identity

$$\boldsymbol{l}_I^1(E) = \boldsymbol{l}_I^1\{E\}$$

is proved. At the same time we have shown that the $\boldsymbol{\varepsilon}$-topology on $\boldsymbol{l}_I^1(E)$ is finer than the $\boldsymbol{\pi}$-topology so that the two topologies coincide.

## 7.3. A Characterization of Nuclear Locally Convex Spaces

**7.3.1.** As an immediate consequence of Theorem 4.2.4 and Proposition 7.2.6 we obtain the

**Theorem.** *A locally convex space $E$ is nuclear if and only if for some, resp. each, infinite index set $I$ the identity*

$$\boldsymbol{l}_I^1 \otimes_\varepsilon E \equiv \boldsymbol{l}_I^1 \otimes_\pi E$$

*holds.*

**7.3.2.** As an important supplement to this characterization of nuclear locally convex spaces we obtain the

**Proposition.** *If $E$ is a nuclear locally convex space, then for each arbitrary locally convex space $F$ the identity*

$$E \otimes_\varepsilon F \equiv E \otimes_\pi F$$

*is valid.*

*Proof.* For an arbitrary given zero neighborhood $U \in \mathfrak{U}(E)$ we determine a zero neighborhood $V \in \mathfrak{U}(E)$ with $V < U$ for which the canonical mapping of $E(V)$ onto $E(U)$ is nuclear. Then there are linear forms $a_n \in E'$ and elements $x_n \in E$ with

$$\varrho = \sum_N p_{V^\circ}(a_n)\, p_U(x_n) < +\infty$$

so that the relation

$$\lim_{\mathfrak{n}} p_U\left(x - \sum_{\mathfrak{n}} \langle x, a_n\rangle\, x_n\right) = 0 \qquad (*)$$

holds for all $x \in E$.

Now if $W$ is an arbitrary zero neighborhood from $\mathfrak{U}(F)$, there exists by the Hahn Banach Theorem for each element

$$z_0 = \sum_{r=1}^{s} x_r^0 \otimes y_r^0 \in E \otimes F$$

in $E \otimes_\pi F$ a continuous linear form $c$ with

$$\langle z_0, c\rangle = \boldsymbol{\pi}_{(U,W)}(z_0) \quad \text{and} \quad |\langle z, c\rangle| \leqq \boldsymbol{\pi}_{(U,W)}(z) \quad \text{for all } z \in E \otimes F .$$

In particular we have the inequality

$$|\langle x \otimes y, c\rangle| \leqq \boldsymbol{\pi}_{(U,W)}(x \otimes y) \leqq p_U(x)\, p_W(y) ,$$

for all elements $x \in E$ and $y \in F$, and from (*) we obtain the identity

$$\langle x \otimes y, c\rangle = \sum_N \langle x, a_n\rangle \langle x_n \otimes y, c\rangle .$$

Consequently, we have the relation

$$\boldsymbol{\pi}_{(U,W)}(z_0) = \langle z_0, c\rangle = \sum_{r=1}^{s} \langle x_r^0 \otimes y_r^0, c\rangle = \sum_N \sum_{r=1}^{s} \langle x_r^0, a_n\rangle \langle x_n \otimes y_r^0, c\rangle$$

$$= \sum_N \left\langle x_n \otimes \sum_{r=1}^{s} \langle x_r^0, a_n\rangle\, y_r^0, c\right\rangle ,$$

from which we get the estimate

$$\boldsymbol{\pi}_{(U,W)}(z_0) \leqq \sum_N p_U(x_n)\, p_W\left(\sum_{r=1}^{s} \langle x_r^0, a_n\rangle\, y_r^0\right) .$$

Finally, it follows from the definition of the semi-norm $\boldsymbol{\varepsilon}_{(V,W)}(z_0)$ that for all linear forms $a \in E'(V^0)$ the inequality

$$p_W\left(\sum_{r=1}^{s} \langle x_r^0, a\rangle\, y_r^0\right) \leqq p_{V^0}(a)\, \boldsymbol{\varepsilon}_{(V,W)}(z_0)$$

is valid. We thus have

$$\boldsymbol{\pi}_{(U,W)}(z_0) \leqq \sum_N p_U(x_n)\, p_{V^0}(a_n)\, \boldsymbol{\varepsilon}_{(V,W)}(z_0) .$$

But since $z_0$ was an arbitrary element from $E \otimes F$, the inequality

$$\boldsymbol{\pi}_{(U,W)}(z) \leqq \varrho\, \boldsymbol{\varepsilon}_{(V,W)}(z) \qquad \text{for all } z \in E \otimes F$$

is proved. Therefore, the $\boldsymbol{\varepsilon}$-topology is finer than the $\boldsymbol{\pi}$-topology on the algebraic tensor product $E \otimes F$ so that the two topologies coincide because of Proposition 7.1.3.

**7.3.3.** By combining Theorem 7.3.1 and Proposition 7.3.2 we obtain the

**Theorem.** *A locally convex space E is nuclear if and only if for each arbitrary locally convex space F the identity*

$$E \otimes_\varepsilon F \equiv E \otimes_\pi F$$

*holds.*

**7.3.4.** One might ask what properties a locally convex space $F$ must have in order that the identity

$$E \otimes_\varepsilon F \equiv E \otimes_\pi F$$

hold only for nuclear locally convex spaces $E$. It is clear that $F$ itself cannot be nuclear. It is not yet known whether this assumption is also sufficient.

In particular we encounter the following

**Problem.** *Is a locally convex space nuclear if the identity*

$$E \otimes_\varepsilon E \equiv E \otimes_\pi E$$

*holds?*

## 7.4. The Kernel Theorem

**7.4.1.** In the sequel we denote by $\mathscr{B}(E, F)$ the set of all continuous bilinear forms defined on two locally convex spaces $E$ and $F$. It is known that a bilinear form $B = B(x, y)$ is continuous if and only if the inequality

$$|B(x, y)| \leqq p_U(x)\, p_V(y) \qquad \text{for all} \quad x \in E \quad \text{and} \quad y \in F$$

is valid for two zero neighborhoods $U \in \mathfrak{U}(E)$ and $V \in \mathfrak{U}(F)$.

**7.4.2.** A bilinear form $B$ is called **nuclear** if it can be represented in the form

$$B(x, y) = \sum_N \langle x, a_n\rangle \langle y, b_n\rangle \qquad \text{for} \quad x \in E \quad \text{and} \quad y \in F\,.$$

Here the linear forms $a_n \in E'$ and $b_n \in F'$ are assumed to satisfy the inequality

$$\sum_N p_{U^0}(a_n)\, p_{V^0}(b_n) < +\infty$$

with two zero neighborhoods $U \in \mathfrak{U}(E)$ and $V \in \mathfrak{U}(F)$. All nuclear bilinear forms are continuous.

**7.4.3.** After these preliminaries we obtain the

**Kernel Theorem.** *If $E$ is a nuclear locally convex space, then for each arbitrary locally convex space $F$ all bilinear forms $B \in \mathscr{B}(E, F)$ are nuclear.*

*Proof.* We determine two zero neighborhoods $U \in \mathfrak{U}(E)$ and $W \in \mathfrak{U}(F)$ such that the inequality

$$|B(x, y)| \leqq p_U(x)\, p_W(y) \quad \text{for all } x \in E \text{ and } y \in F$$

is valid. Since the locally convex space $E$ is supposed to be nuclear there is a zero neighborhood $V \in \mathfrak{U}(E)$ with $V < U$ such that the canonical mapping from $E(V)$ onto $E(U)$ is nuclear. Thus there exist linear forms $a_n \in E'$ and elements $x_n \in E$ with

$$\sum_N p_{V^0}(a_n)\, p_U(x_n) < +\infty$$

and

$$\lim_{\mathfrak{n}} p_U\left((x - \sum_{\mathfrak{n}} \langle x, a_n \rangle x_n\right) = 0 \quad \text{for } x \in E .$$

We now obtain the representation

$$B(x, y) = \sum_N \langle x, a_n \rangle B(x_n, y) \quad \text{with } x \in E \text{ and } y \in F$$

for the bilinear form $B$. If we use the linear forms $b_n \in F'$ defined by

$$\langle y, b_n \rangle = B(x_n, y) \quad \text{for } y \in F$$

we then have

$$p_{W^0}(b_n) = \sup \{|\langle y, b_n \rangle| : y \in W\} \leqq p_U(x_n) .$$

Consequently, the inequality

$$\sum_N p_{V^0}(a_n)\, p_{W^0}(b_n) \leqq \sum_N p_{V^0}(a_n)\, p_U(x_n) < +\infty$$

holds. Moreover, we have

$$B(x, y) = \sum_N \langle x, a_n \rangle \langle y, b_n \rangle \quad \text{for all } x \in E \text{ and } y \in F .$$

**7.4.4.** The property of nuclear locally convex spaces presented in 7.4.3 also characterizes them. In fact, we have the

**Theorem.** *A locally convex space $E$ is nuclear if and only if for each arbitrary locally convex space $F$, all bilinear forms $B \in \mathcal{B}(E, F)$ are nuclear.*

*Proof.* Since the necessity of the condition has already been proved, we have only to show its sufficiency. For this purpose we consider an arbitrary zero neighborhood $U \in \mathfrak{U}(E)$ and define a bilinear from $B \in \mathcal{B}(E, E'(U^0))$ by the equation

$$B(x, a) = \langle x, a \rangle \quad \text{with } x \in E \text{ and } a \in E'(U^0) .$$

By hypothesis, this bilinear form can be represented in the form

$$\langle x, a \rangle = \sum_N \langle x, a_n \rangle \langle \mathfrak{x}_n, a \rangle ,$$

where the linear forms $a_n \in E'$ and $\mathfrak{x}_n \in E'(U^0)'$ satisfy the inequality

$$\sum_N p_{V^0}(a_n)\, p'_{U^0}(\mathfrak{x}_n) < +\infty$$

for some zero neighborhood $V \in \mathfrak{U}(E)$. Thus for $x \in E$ and $a \in U^0$ we have the estimate

$$|\langle x, a\rangle| \leqq \sum_N |\langle x, a_n\rangle|\, p'_{U^0}(\mathfrak{x}_n)\,,$$

from which we obtain the relation

$$p_U(x) \leqq \sum_N |\langle x, b_n\rangle| \quad \text{for} \quad x \in E$$

with the linear forms $b_n = p'_{U^0}(x_n)\, a_n \in E'$. However, since the inequality

$$\sum_N p_{V^0}(b_n) = \sum_N p_{V^0}(a_n)\, p'_{U^0}(\mathfrak{x}_n) < +\infty$$

holds, the locally convex space $E$ must be nuclear by Proposition 4.1.4.

**7.4.5.** By analogy to 7.3.4, we have the following

**Problem.** *Is a locally convex space $E$ nuclear, if all bilinear forms $B \in \mathscr{B}(E, E)$ are nuclear?*

A. Grothendieck has already shown ([3], Ch. II, p. 47) that the questions stated in Problems 7.3.4 and 7.4.5 can be answered in the affirmative if the topology of the locally convex space $E$ can be constructed from a system of semi-scalar products. This hypothesis is always satisfied for nuclear locally convex spaces because of Theorem 4.4.2. An elementary proof of this fact was given by K. John [1].

## 7.5. The Complete $\pi$-Tensor Product of Normed Spaces

**7.5.1.** Let $E$ and $F$ be two normed spaces with closed unit balls $U$ and $V$. We obtain the locally convex $\pi$-topology on the algebraic tensor product $E \otimes F$ from the norm $\pi_{(U,V)}$, and have the

**Theorem.** *Each element $z$ of the complete tensor product $E \tilde{\otimes}_\pi F$ can be represented in the form*

$$z = \sum_N x_n \otimes y_n \quad \textit{with} \quad x_n \in E \quad \textit{and} \quad y_n \in F$$

*such that for an arbitrary given positive number $\delta$ the inequality*

$$\sum_N p_U(x_n)\, p_V(y_n) \leqq \pi_{(U,V)}(z) + \delta$$

*holds.*

*Proof.* Since the element $z$ is in the complete hull of the normed space $E \otimes_\pi F$, there is a sequence of elements $z_r \in E \otimes F$ with

$$\boldsymbol{\pi}_{(U,V)}(z - z_r) < \frac{1}{2^{r+3}}\,\delta \quad \text{for} \quad r \in R = \{0, 1, 2, \ldots\}\,.$$

From the inequality

$$\boldsymbol{\pi}_{(U,V)}(z_{r+1} - z_r) < \frac{1}{2^{r+2}}\,\delta \quad \text{for} \quad r \in R\,,$$

it follows that the element $z_{r+1} - z_r$ can be represented as

$$z_{r+1} - z_r = \sum_{n=1}^{n_{r+1}} x_n^{[r+1]} \otimes y_n^{[r+1]}$$

with

$$x_n^{[r+1]} \in E\,, \quad y_n^{[r+1]} \in F \quad \text{and} \quad \sum_{n=1}^{n_{r+1}} p_U(x_n^{[r+1]})\, p_V(y_n^{[r+1]}) \leqq \frac{1}{2^{r+2}}\,\delta\,.$$

Since

$$\boldsymbol{\pi}_{(U,V)}(z_0) \leqq \boldsymbol{\pi}_{(U,V)}(z) + \boldsymbol{\pi}_{(U,V)}(z - z_0) < \boldsymbol{\pi}_{(U,V)}(z) + \frac{1}{2}\,\delta$$

we can write the element $z_0$ as

$$z_0 = \sum_{n=1}^{n_0} x_n^{[0]} \otimes y_n^{[0]}$$

with

$$x_n^{[0]} \in E\,, \quad y_n^{[0]} \in F \quad \text{and} \quad \sum_{n=1}^{n_0} p_U(x_n^{[0]})\, p_V(y_n^{[0]}) \leqq \boldsymbol{\pi}_{(U,V)}(z) + \frac{1}{2}\,\delta\,.$$

Consequently, we have

$$z = z_0 + \sum_R (z_{r+1} - z_r) = \sum_R \sum_{n=1}^{n_r} x_n^{[r]} \otimes y_n^{[r]}\,.$$

Our assertion is thus completely proved because of the inequality

$$\sum_R \sum_{n=1}^{n_r} p_U(x_n^{[r]})\, p_V(y_n^{[r]}) \leqq \boldsymbol{\pi}_{(U,V)}(z) + \delta\,.$$

**7.5.2.** If we associate with each element

$$z = \sum_{r=1}^{n} a_r \otimes y_r \in E' \otimes F$$

the mapping

$$T\,x = \sum_{r=1}^{n} \langle x, a_r \rangle\, y_r\,,$$

then the algebraic tensor product $E' \otimes F$ can be identified with the linear space $\mathcal{A}(E, F)$. In particular, if we choose the representation of the element $z$ in such a way that the inequality

$$\sum_{r=1}^{n} p_{U^\circ}(a_r)\, p_V(y_r) \leqq \boldsymbol{\pi}_{(U^\circ,V)}(z) + \delta$$

holds for a given positive number $\delta$, we then get the estimate

$$\beta(T) \leqq \pi_{(U^0, V)}(z) + \delta ,$$

from which we obtain the relation

$$\beta(T) \leqq \pi_{(U^0, V)}(z)$$

by taking the limit as $\delta \to 0$. Consequently, the canonical mapping of $E'_b \otimes_\pi F$ into $\mathcal{L}_b(E, F)$ is continuous. It can therefore be uniquely extended to a continuous linear mapping of $E'_b \tilde{\otimes}_\pi F$ into $\mathcal{L}_b(E, F)$. Because of Theorem 7.5.1 we obtain in this way set the of all nuclear mappings from $E$ into $F$ as the images of elements of $E'_b \tilde{\otimes}_\pi F$.

For a large class of normed spaces it can even be shown that the canonical mapping of $E'_b \tilde{\otimes}_\pi F$ onto $\mathcal{N}(E, \tilde{F})$ is one-to-one. The question of whether this assertion is true for arbitrary normed spaces is still unanswered.

Chapter 8

# Mappings of Type $l^p$ and s

Let $E$ and $F$ be two normed spaces. For each mapping $T \in \mathcal{L}(E, F)$, the approximation numbers $\alpha_r(T)$ with $r = 0, 1, \ldots$, which are defined in 8.1.1, provide a measure of how well $T$ can be approximated by finite mappings whose range is at most $r$-dimensional. For completely continuous mappings in Hilbert spaces the approximation numbers coincide with eigenvalues (ordered according to magnitude) of the mappings $|T| = (T^* T)^{1/2}$ (Theorem 8.3.2).

We say that a mapping $T \in \mathcal{L}(E, F)$ is of *type* $\boldsymbol{l}^p$ if the inequality

$$\sum_R \alpha_r(T)^p < +\infty$$

holds. Mappings of this type form a linear space. Products of these mappings are studied in 8.2.7.

Our most important result is the assertion that each mapping of type $\boldsymbol{l}^1$ is nuclear (Theorem 8.4.3). In Hilbert spaces the converse of this statement is also valid. However, for arbitrary Banach spaces it only can be shown that the product of two nuclear mappings is of type $\boldsymbol{l}^2$ (Theorem 8.4.5).

A class of mappings which can be very well approximated is the class of mappings of *type* **s** which coincides with the class of *Fredholm mappings of order* 0 introduced by A. Grothendieck ([3], Chap. II, p. 6) (Theorem 8.5.6).

In the last section, mappings of type $\boldsymbol{l}^p$ and **s** will be used to characterize nuclear locally convex spaces.

The results of this chapter are essentially due to A. Pietsch [5].

## 8.1. The Approximation Numbers of Continuous Linear Mappings in Normed Spaces

**8.1.1.** For two arbitrary normed spaces $E$ and $F$, $\mathcal{A}_r(E, F)$ for $r = 0, 1, \ldots$ denotes the collection of all finite mappings $A \in \mathcal{L}(E, F)$ whose range $\mathfrak{R}(A)$ is at most $r$-dimensional.

For an arbitrary mapping $T \in \mathcal{L}(E, F)$ we designate

$$\alpha_r(T) = \inf \{\boldsymbol{\beta}(T - A) : A \in \mathcal{A}_r(E, F)\} \qquad (r = 0, 1, \ldots)$$

as the **r-th approximation number** of $T$. Clearly, we always have

$$\beta(T) = \alpha_0(T) \geqq \alpha_1(T) \geqq \alpha_2(T) \geqq \cdots \geqq 0 .$$

**8.1.2.** In the sequel we list some elementary properties of approximation numbers.

**Proposition 1.** *For two mappings $S, T \in \mathcal{L}(E, F)$ we always have*

$$\alpha_{r+s}(S + T) \leqq \alpha_r(S) + \alpha_s(T) .$$

*Proof.* For an arbitrary positive number $\delta$ we determine mappings $A \in \mathcal{A}_r(E, F)$ and $B \in \mathcal{A}_s(E, F)$ with

$$\beta(S - A) \leqq \alpha_r(S) + \delta \quad \text{and} \quad \beta(T - B) \leqq \alpha_s(T) + \delta .$$

Then since $A + B \in \mathcal{A}_{r+s}(E, F)$ the inequality

$$\alpha_{r+s}(S + T) \leqq \beta(S + T - A - B) \leqq \beta(S - A) + \beta(T - B) \leqq \alpha_r(S) + \alpha_s(T) + 2\,\delta$$

is valid, an d by taking the limit as $\delta \to 0$ we get the stated inequality.

**Proposition 2.** *For two mappings $S, T \in \mathcal{L}(E, F)$ we always have*

$$|\alpha_r(S) - \alpha_r(T)| \leqq \beta(S - T) .$$

*Proof.* Because of Proposition 1 we have

$$\alpha_r(S) \leqq \alpha_r(T) + \alpha_0(S - T) \quad \text{or} \quad \alpha_r(S) - \alpha_r(T) \leqq \beta(S - T) .$$

By interchanging $S$ and $T$ we get the inequality

$$|\alpha_r(S) - \alpha_r(T)| \leqq \beta(S - T) .$$

Without proof we formulate

**Proposition 3.** *For each mapping $T \in \mathcal{L}(E, F)$ and all numbers $\lambda$ we have*

$$\alpha_r(\lambda\, T) = |\lambda|\, \alpha_r(T) .$$

**Proposition 4.** *For each mapping $T \in \mathcal{L}(E, F)$, $\alpha_r(T) = 0$ always implies $T \in \mathcal{A}_r(E, F)$.*

*Proof.* We assume that the dimension of the range of $T$ is larger than $r$. There are then $r + 1$ linearly independent elements $y_i = T\, x_i$ for which we can determine $r + 1$ linear forms $b_k \in F'$ with $\langle y_i, b_k\rangle = \delta_{ik}$.

Since $\det\, \{\delta_{ik}\} = 1$, there exists a positive number $\sigma$ such that

$$\det\, \{\alpha_{ik}\} \neq 0 \quad \text{for} \quad |\delta_{ik} - \alpha_{ik}| \leqq \sigma .$$

Since the relation

$$\inf\, \{\beta(T - A) : A \in \mathcal{A}_r(E, F)\} = 0$$

holds by hypothesis, there is for the positive number

$$\varrho = \sigma \max \{p_U(x_i)\, p_{V^\circ}(b_k) : i, k = 1, \ldots, r+1\}^{-1}$$

a mapping $A \in \mathcal{A}_r(E, F)$ with $\beta(T - A) \leqq \varrho$. Since

$$|\delta_{ik} - \langle A x_i, b_k \rangle| = |\langle T\, x_i - A\, x_i, b_k \rangle| \leqq \varrho\, p_U(x_i)\, p_{V^\circ}(b_k) \leqq \sigma$$

we have the assertion

$$\det \{\langle A\, x_i, b_k \rangle\} \neq 0 .$$

But since the range of $A$ is at most $r$-dimensional, the $r+1$ elements $A x_i$ are linearly independent. We thus have

$$\det \{\langle A\, x_i, b_k \rangle\} = 0 .$$

From the contradiction thus encountered it follows that the assumption $\dim \mathfrak{R}(T) > r$ is false. Thus $T \in \mathcal{A}_r(E, F)$.

For $E$, $F$ and $G$ three arbitrary normed spaces, we obtain

**Proposition 5.** *For two mappings $T \in \mathcal{L}(E, F)$ and $S \in \mathcal{L}(F, G)$ we have*

$$\alpha_{r+s}(S\, T) \leqq \alpha_r(S)\, \alpha_s(T) .$$

*Proof.* For an arbitrary positive number $\delta$ we determine mappings $B \in \mathcal{A}_s(E, F)$ and $A \in \mathcal{A}_r(F, G)$ with

$$\beta(T - B) \leqq \alpha_s(T) + \delta \quad \text{and} \quad \beta(S - A) \leqq \alpha_r(S) + \delta .$$

Then since $A(T - B) + S\, B \in \mathcal{A}_{r+s}(E, G)$, we have the estimate

$$\begin{aligned} \alpha_{r+s}(S\, T) &\leqq \beta(S\, T - A\,(T - B) - S\, B) \leqq \beta(S - A)\, \beta(T - B) \\ &\leqq (\alpha_r(S) + \delta)\, (\alpha_s(T) + \delta) , \end{aligned}$$

from which we get the required inequality by taking the limit as $\delta \to 0$.

**8.1.3.** If $F$ is a linear subspace of the normed space $G$, then each continuous linear mapping $T$ from $E$ into $F$ can be considered as a mapping from $E$ into $G$. Thus $T$ has approximation numbers in $\mathcal{L}(E, G)$ as well as in $\mathcal{L}(E, F)$, which we will distinguish by $\alpha_r^G(T)$, resp. $\alpha_r^F(T)$. It is easy to see that the inequality

$$\alpha_r^G(T) \leqq \alpha_r^F(T)$$

holds.

**Proposition.** *If $F$ is a dense linear subspace of the normed space $G$, then for each continuous linear mapping $T$ of a normed space $E$ into $F$ we have the identity*

$$\alpha_r^F(T) = \alpha_r^G(T) .$$

*Proof.* For $\delta$ an arbitrary positive number we determine a mapping $B \in \mathcal{A}_r(E, G)$ with

$$\boldsymbol{\beta}(T - B) \leqq \alpha_r^G(T) + \delta .$$

Since the mapping $B$ can be represented as

$$B\,x = \sum_{n=1}^{r} \langle x, a_n\rangle\, z_n \qquad \text{for} \quad x \in E$$

with linear forms $a_1, \ldots, a_r \in E'$ and elements $z_1, \ldots, z_r \in G$, there are elements $y_1, \ldots, y_r \in F$ with

$$\sum_{n=1}^{r} p_{U^\circ}(a_n)\, p_W(z_n - y_n) \leqq \delta ,$$

and we can consider the mapping $A \in \mathcal{A}_r(E, F)$ with

$$A\,x = \sum_{n=1}^{r} \langle x, a_n\rangle\, y_n \qquad \text{for} \quad x \in E .$$

Then we have

$$\boldsymbol{\beta}(B - A) \leqq \sum_{n=1}^{r} p_{U^\circ}(a_n)\, p_W(z_n - y_n) \leqq \delta .$$

Consequently the estimate

$$\alpha_r^F(T) \leqq \boldsymbol{\beta}(T - A) \leqq \boldsymbol{\beta}(T - B) + \boldsymbol{\beta}(B - A) \leqq \alpha_r^G(T) + 2\,\delta$$

holds, and by taking the limit as $\delta \to 0$ we get the inequality

$$\alpha_r^F(T) \leqq \alpha_r^G(T) .$$

Our assertion is thus proved since, as we have just remarked,

$$\alpha_r^G(T) \leqq \alpha_r^F(T)$$

is always true.

**8.1.4.** We now give a procedure for determining approximation numbers. For this we need the following

**Lemma.** *Let $T$ be a continuous linear mapping from an arbitrary normed space $E$ into an $(r+1)$-dimensional normed space $F$ for which there is a mapping $S \in \mathcal{L}(F, E)$ with $T\,S\,y = y$ for $y \in F$. We then have the inequality*

$$\alpha_r(T)\, \boldsymbol{\beta}(S) \geqq 1 .$$

*Proof.* Assume that

$$\alpha_r(T)\, \boldsymbol{\beta}(S) < 1 .$$

Then there is a mapping $A \in \mathcal{A}_r(E, F)$ with

$$\boldsymbol{\beta}(T - A)\, \boldsymbol{\beta}(S) < 1 .$$

Since $F$ is also a Banach space, with identity mapping denoted by $I$, $I - (T - A) S = A S$ must be invertable. But this is impossible since $A S \in \mathcal{A}_r(F, F)$. The validity of our assertion follows from this contradiction.

**8.1.5.** As an application of the previous lemma we calculate the approximation numbers of a simple mapping. For this purpose we shall denote by $\mathfrak{F}_r(I)$ the collection of all $r$-element subsets of an arbitrary index set $I$.

**Proposition.** *For each mapping $T \in \mathfrak{L}(\boldsymbol{m}_I, \boldsymbol{m}_I)$ of the form*

$$T[\xi_i, I] = [\tau_i \xi_i, I] \qquad \textit{with} \quad [\tau_i, I] \in \boldsymbol{m}_I$$

*we have the identity*

$$\alpha_r(T) = \sup\{\inf\{|\tau_i| : i \in \mathfrak{i}\} : \mathfrak{i} \in \mathfrak{F}_{r+1}(I)\} .$$

*Proof.* We denote the expression on the right side of the stated identity $\sigma_r$. Then the set

$$\mathfrak{i}_0 = \{i \in I : |\tau_i| > \sigma_r\}$$

has at most $r$ elements, and the expression

$$A[\xi_i, I] = [\tau_i(\mathfrak{i}_0)\, \xi_i, I]$$

defines a mapping $A$ in $\mathcal{A}_r(\boldsymbol{m}_I, \boldsymbol{m}_I)$. Consequently, we have

$$\alpha_r(T) \leqq \boldsymbol{\beta}(T - A) = \sup\{|\tau_i| : i \in I \backslash \mathfrak{i}_0\} \leqq \sigma_r .$$

For the case $\sigma_r = 0$, our assertion is thus already proved. Otherwise, we consider an arbitrary set $\mathfrak{i} \in \mathfrak{F}_{r+1}(I)$ with

$$\varrho_{\mathfrak{i}} = \inf\{|\tau_i| : i \in \mathfrak{i}\} > 0$$

and define a mapping $P_{\mathfrak{i}} \in \mathfrak{L}(\boldsymbol{m}_I, \boldsymbol{m}_{\mathfrak{i}})$ with $\boldsymbol{\beta}(P_{\mathfrak{i}}) = 1$ by

$$P_{\mathfrak{i}}[\xi_i, I] = [\xi_i, \mathfrak{i}] .$$

Since the mapping $T_{\mathfrak{i}} = P_{\mathfrak{i}}\, T$ has the form

$$T_{\mathfrak{i}}[\xi_i, I] = [\tau_i \xi_i, \mathfrak{i}]$$

we can determine a mapping $S_{\mathfrak{i}} \in \mathfrak{L}(\boldsymbol{m}_{\mathfrak{i}}, \boldsymbol{m}_I)$ by the formula

$$S_{\mathfrak{i}}[\xi_i, \mathfrak{i}] = [\eta_i, I] \quad \text{with} \quad \eta_i = \begin{cases} \tau_i^{-1} \xi_i & \text{for } i \in \mathfrak{i} \\ 0 & \text{for } i \notin \mathfrak{i} \end{cases}$$

with $\boldsymbol{\beta}(S_{\mathfrak{i}}) = \varrho_{\mathfrak{i}}^{-1}$, for which

$$T_{\mathfrak{i}}\, S_{\mathfrak{i}}[\xi_i, \mathfrak{i}] = [\xi_i, \mathfrak{i}]$$

always holds with $[\xi_i, \mathfrak{i}] \in \boldsymbol{m}_{\mathfrak{i}}$. Consequently, we have

$$\varrho_{\mathfrak{i}} = \boldsymbol{\beta}(S_{\mathfrak{i}})^{-1} \leqq \alpha_r(T_{\mathfrak{i}}) \leqq \boldsymbol{\beta}(P_{\mathfrak{i}})\, \alpha_r(T) = \alpha_r(T) .$$

Since $\mathfrak{i}$ was an arbitrary subset from $\mathfrak{F}_{r+1}(I)$ we also have

$$\sigma_r \leqq \alpha_r(T) ,$$

and the identity is proved.

## 8.2. Mappings of Type $l^p$

**8.2.1.** For two arbitrary normed spaces $E$ and $F$ we consider for each positive number $p$ the collection $\boldsymbol{l}^p(E, F)$ of all mappings $T \in \mathcal{L}(E, F)$ for which

$$\sum_R \alpha_r(T)^p < +\infty \quad \text{with} \quad R = \{0, 1, 2, \ldots\}$$

is valid, and we say, these mappings are of **type $\boldsymbol{l}^p$**.

**8.2.2. Proposition.** *$\boldsymbol{l}^p(E, F)$ is a linear space.*

*Proof.* Since for two numbers $\xi, \eta \geqq 0$ the inequality

$$(\xi + \eta)^p \leqq \tau_p (\xi^p + \eta^p) \quad \text{with} \quad \tau^p = \max\{2^{p-1}, 1\}$$

holds, we obtain the estimate

$$\sum_R \alpha_r(S + T)^p \leqq 2 \sum_R \alpha_{2r}(S + T)^p$$
$$\leqq 2 \sum_R [\alpha_r(S) + \alpha_r(T)]^p \leqq 2 \tau_p \sum_R [\alpha_r(S)^p + \alpha_r(T)^p] ,$$

for two mappings $S, T \in \boldsymbol{l}^p(E, F)$ by using Proposition 1 from 8.1.2. Consequently, the mapping $S + T$ is of type $\boldsymbol{l}^p$.

On the other hand, since

$$\sum_R \alpha_r(\lambda T)^p = |\lambda|^p \sum_R \alpha_r(T)^p$$

for each number $\lambda$, we have $\lambda T \in \boldsymbol{l}^p(E, F)$ whenever $T \in \boldsymbol{l}^p(E, F)$.

**8.2.3.** By

$$\varrho_p(T) = \left\{\sum_R \alpha_r(T)^p\right\}^{1/p}$$

we define on $\boldsymbol{l}^p(E, F)$ a real valued function with the following properties.

($Q_1$) $\varrho_p(T) \geqq 0$.
($Q_2$) *From* $\varrho_p(T) = 0$ *it follows that* $T = 0$.
($Q_3$) *For all numbers* $\lambda$ *we have* $\varrho_p(\lambda T) = |\lambda|\, \varrho_p(T)$.
($Q_4$) *For some number* $\sigma_p \geqq 1$ *we have the inequality*

$$\varrho_p(S + T) \leqq \sigma_p[\varrho_p(S) + \varrho_p(T)] \quad \textit{for} \quad S, T \in \boldsymbol{l}^p(E, F) .$$

Since the statements $(Q_1)$, $(Q_2)$ and $(Q_3)$ are immediately clear, we shall only prove the inequality in $(Q_4)$. For this we write the estimate derived in 8.2.2 in the form

$$\varrho_p(S+T)^p \leqq 2\tau_p[\varrho_p(S)^p + \varrho_p(T)^p] .$$

Consequently, we have

$$\varrho_p(S+T) \leqq (2\tau_p)^{1/p}\tau_{1/p}[\varrho_p(S) + \varrho_p(T)] ,$$

and we can use the number

$$(2\tau_p)^{1/p}\tau_{1/p} = \begin{cases} 2 & \text{for } p \geqq 1 \\ 2^{2/p-1} & \text{for } p \leqq 1 \end{cases}$$

for $\sigma_p$.

On the basis of the properties stated, $\varrho_p(T)$ will be designated as a **quasi-norm.** We obtain a metric topology on $\mathfrak{l}^p(E, F)$ by using the sets

$$U_\varepsilon(T) = \{S \in \mathfrak{l}^p(E, F) : \varrho_p(S-T) \leqq \varepsilon\} \quad \text{with } \varepsilon > 0$$

as a fundamental system of neighborhoods of the mapping $T$.

**8.2.4. Lemma.** *If $\{T_\alpha\}$ is a directed Cauchy-system from $\mathfrak{l}^p(E, F)$ and there is a mapping $T \in \mathfrak{L}(E, F)$ with $\lim\limits_\alpha T_\alpha x = T x$ for $x \in E$, then $T$ belongs to $\mathfrak{l}^p(E, F)$ and we have $\varrho_p - \lim\limits_\alpha T_\alpha = T$.*

*Proof.* Since we have the inequality $\varrho_p(S) \geqq \beta(S)$ for all mappings $S \in \mathfrak{l}^p(E, F)$, $\{T_\alpha\}$ is also a directed Cauchy-system in $\mathfrak{L}(E, F)$ which converges to the mapping $T$ because $\lim\limits_\alpha T_\alpha x = T x$ for $x \in E$.

According to Proposition 2 from 8.1.2 we have the estimate

$$|\alpha_r(T - T_\alpha) - \alpha_r(T_\beta - T_\alpha)| \leqq \beta(T - T_\beta)$$

so that

$$\lim_\beta \alpha_r(T_\beta - T_\alpha) = \alpha_r(T - T_\alpha) .$$

For an arbitrary given positive number $\delta$ we determine $\alpha_0$ so that

$$\varrho_p(T_\beta - T_\alpha) = \left\{\sum_R \alpha_r(T_\beta - T_\alpha)^p\right\}^{1/p} \leqq \delta \quad \text{for } \alpha, \beta \geqq \alpha_0 .$$

Then by taking the limit we get the equation

$$\varrho_p(T - T_\alpha) = \left\{\sum_R \alpha_r(T - T_\alpha)^p\right\}^{1/p} \leqq \delta \quad \text{for } \alpha \geqq \alpha_0 .$$

Consequently, the mapping $T - T_{\alpha_0}$ and thus $T$, as well, belongs to $\mathfrak{l}^p(E, F)$ and the directed Cauchy-system $\{T_\alpha\}$ converges to $T \in \mathfrak{l}^p(E, F)$.

As a simple consequence of the lemma just proved we get the

**Proposition.** *For a normed space $E$ and a Banach space $F$, $\mathfrak{l}^p(E, F)$ is complete.*

**8.2.5. Proposition.** *The linear space $\mathcal{A}(E, F)$ is dense in $\mathfrak{l}^p(E, F)$.*

*Proof.* Clearly, every finite mapping $T \in \mathcal{A}(E, F)$ belongs to $\mathfrak{l}^p(E, F)$, since the sequence $\{\alpha_r(T)\}$ contains only finitely many numbers different from 0.

On the other hand, if $T$ is a mapping in $\mathfrak{l}^p(E, F)$, there is for each positive number $\delta$ a natural number $s$ with

$$\sum_{r=s}^{\infty} \alpha_r(T)^p < \frac{1}{4}\delta^p .$$

Since the sequence of approximation numbers decreases monotonically we have

$$s\,\alpha_{2s}(T)^p \leqq \sum_{r=s+1}^{2s} \alpha_r(T)^p < \frac{1}{4}\delta^p .$$

We now determine a mapping $A \in \mathcal{A}_{2s}(E, F)$ with

$$s\,\beta(T - A)^p < \frac{1}{4}\delta^p .$$

We then have

$$\alpha_{r+2s}(T - A) \leqq \alpha_r(T) \qquad \text{for } r = 0, 1, \ldots,$$

and arrive at the inequality

$$\begin{aligned} \varrho_p(T - A)^p &= \sum_{r=0}^{3s-1} \alpha_r(T - A)^p + \sum_{r=3s}^{\infty} \alpha_r(T - A)^p \\ &\leqq 3\,s\,\beta(T - A)^p + \sum_{r=s}^{\infty} \alpha_r(T)^p \leqq \frac{3}{4}\delta^p + \frac{1}{4}\delta^p = \delta^p . \end{aligned}$$

Thus our assertion is proved because we have determined for the arbitrary positive number $\delta$ a mapping $A \in \mathcal{A}(E, F)$ with $\varrho_p(T - A) \leqq \delta$.

**8.2.6. Proposition.** *Every mapping of type $\mathfrak{l}^p$ is precompact.*

*Proof.* By 8.2.5 there is for each mapping $T \in \mathfrak{l}^p(E, F)$ a sequence of mappings $T_n \in \mathcal{A}(E, F)$ with

$$\lim_n \varrho_p(T - T_n) = 0 .$$

However, since the inequality $\varrho_p(S) \geqq \beta(S)$ holds for all mappings $S \in \mathfrak{l}^p(E, F)$, we have

$$\lim_n \beta(T - T_n) = 0 .$$

Consequently the mapping $T$ is precompact by 0.10.6.

**8.2.7.** For three normed spaces $E$, $F$ and $G$ we have the important

**Theorem.** *For $T \in \mathfrak{l}^p(E, F)$ and $S \in \mathfrak{l}^q(F, G)$ it follows that $S\,T \in \mathfrak{l}^s(E, G)$ with*

$$\frac{1}{s} = \frac{1}{p} + \frac{1}{q} .$$

*Proof.* By an application of the generalized Hölder inequality

$$\left\{\sum_R |\xi_r \eta_r|^s\right\}^{1/s} \leqq \left\{\sum_R |\xi_r|^p\right\}^{1/p} \cdot \left\{\sum_R |\eta_r|^q\right\}^{1/q}$$

and Proposition 5 from 8.1.2 we get the estimate

$$\varrho_s(S\,T) = \left\{\sum_R \alpha_r(S\,T)^s\right\}^{1/s} \leqq \left\{2 \sum_R \alpha_{2r}(S\,T)^s\right\}^{1/s} \leqq \left\{2 \sum_R [\alpha_r(S)\,\alpha_r(T)]^s\right\}^{1/s}$$

$$\leqq 2^{1/s} \left\{\sum_R \alpha_r(S)^q\right\}^{1/q} \cdot \left\{\sum_R \alpha_r(T)^p\right\}^{1/p} = 2^{1/s}\,\varrho_q(S)\,\varrho_p(T)\,.$$

Therefore, the product $S\,T$ belongs to $\boldsymbol{l}^s(E, G)$.

**8.2.8.** Moreover, we obtain the

**Proposition.**

(1) *If* $T \in \mathscr{L}(E, F)$ *and* $S \in \boldsymbol{l}^p(F, G)$ *then* $S\,T \in \boldsymbol{l}^p(E, G)$ *and* $\varrho_p(S\,T) \leqq \varrho_p(S)\,\beta(T)$.

(2) *If* $T \in \boldsymbol{l}^p(E, F)$ *and* $S \in \mathscr{L}(F, G)$ *then* $S\,T \in \boldsymbol{l}^p(E, G)$ *and* $\varrho_p(S\,T) \leqq \beta(S)\,\varrho_p(T)$.

*Proof.* By Proposition 5 from 8.1.2 we have

$$\alpha_r(S\,T) \leqq \alpha_r(S)\,\beta(T) \qquad \text{resp.} \qquad \alpha_r(S\,T) \leqq \beta(S)\,\alpha_r(T)\,.$$

Consequently, the inequality

$$\varrho_p(S\,T) \leqq \varrho_p(S)\,\beta(T) \qquad \text{resp.} \qquad \varrho_p(S\,T) \leqq \beta(S)\,\varrho_p(T)$$

holds. Therefore, our assertion is proved.

**8.2.9.** Finally we present a simple class of mappings of type $\boldsymbol{l}^p$.

**Proposition.** *A mapping* $T \in \mathscr{L}(\boldsymbol{m}_I, \boldsymbol{m}_I)$ *of the form*

$$T[\xi_i, I] = [\tau_i\,\xi_i, I] \quad \textit{with} \quad [\xi_i, I] \in \boldsymbol{m}_I$$

*is of type* $\boldsymbol{l}^p$ *if and only if the relation*

$$\varrho_p(T) = \left\{\sum_I |\tau_i|^p\right\}^{1/p} < +\infty$$

*holds.*

*Proof.* We first assume that the mapping $T$ is of type $\boldsymbol{l}^p$. Then, we of course have

$$\lim_r \alpha_r(T) = 0\,.$$

Consequently, every set

$$\mathfrak{i}_n = \left\{i \in I : |\tau_i| \geqq \frac{1}{n}\right\} \quad \text{with} \quad n \in N$$

must be finite because otherwise the relation

$$\alpha_r(T) \geqq \frac{1}{n} \quad \text{for all} \quad r \in R$$

would be valid. The set

$$I_0 = \{i \in I : |\tau_i| > 0\} = \bigcup_N \mathfrak{i}_n$$

thus has the form

$$I_0 = \{i_0, i_1, \ldots, i_s\} \quad \text{or} \quad I_0 = \{i_0, i_1, i_2, \ldots\}.$$

Here we can also assume that $m \leqq n$ always implies $|\tau_{i_m}| \geqq |\tau_{i_n}|$. Thus by Proposition 8.1.5 the identities

$$\alpha_r(T) = \begin{cases} |\tau_{i_r}| & \text{for } 0 \leqq r \leqq s \\ 0 & \text{for } r > s \end{cases}$$

hold and we have

$$\left\{\sum_I |\tau_i|^p\right\}^{1/p} = \left\{\sum_R \alpha_r(T)^p\right\}^{1/p} = \varrho_p(T) < +\infty.$$

On the other hand, if $T$ is a mapping with

$$\left\{\sum_I |\tau_i|^p\right\}^{1/p} < +\infty,$$

then the sets

$$\mathfrak{i}_n = \left\{i \in I : |\tau_i| \geqq \frac{1}{n}\right\} \quad \text{with} \quad n \in N$$

are also finite and we get the identity

$$\varrho_p(T) = \left\{\sum_R \alpha_r(T)^p\right\}^{1/p} = \left\{\sum_I |\tau_i|^p\right\}^{1/p} < +\infty$$

exactly as in the first part of our proof. Consequently, the mapping $T$ is of type $\boldsymbol{l}^p$.

## 8.3. The Approximation Numbers of Compact Mappings in Hilbert Spaces

**8.3.1.** In this section we will always assume that $E$ uand $F$ are two Hilbert spaces with closed unit balls $U$ and $V$. In this case, the compact linear mappings from $E$ into $F$ can be represented in a very simple way.

**Spectral Decomposition Theorem.** *For each compact mapping $T \in \mathfrak{L}(E, F)$ there are two orthogonal systems $[e_i, I]$ and $[f_i, I]$ in $E$ resp. $F$ as well as a family of numbers $[\lambda_i, I] \in \mathbf{c}_I$ with $\lambda_i > 0$ such that we have*

$$T x = \sum_I \lambda_i (x, e_i) f_i \quad \textit{for all} \quad x \in E$$

*Proof.* We consider the collection of all orthonormal systems of $E$ which consist of those elements $e$ for which there is a positive number $\lambda$ with

$$T^* T e = \lambda^2 e ,$$

and order this collection with respect to set theoretic inclusion. By Zorn's lemma there is a maximal orthonormal system $[e_i, I]$ for which the equalities

$$T^* T e_i = \lambda_i^2 e_i$$

are valid for certain positive numbers $\lambda_i$. Then by

$$P x = x - \sum_I (x, e_i) e_i \quad \text{for } x \in E \tag{*}$$

we define a continuous Projection $P$, whose range $\mathfrak{R}(P)$ consists of all elements $x \in E$ with $(x, e_i) = 0$ for $i \in I$.

We now assume that for the mapping $T_0 = T P$, $\lambda_0 = \boldsymbol{\beta}(T_0) > 0$. Then by the lemma below there is an element $e_0$ in $E$ with

$$T_0^* T_0 e_0 = \lambda_0^2 e_0 \quad \text{and} \quad p_U(e_0) = 1 .$$

Since $T_0^* = P T^*$ we have

$$e_0 = \lambda_0^{-2} P T^* T P e_0 \in \mathfrak{R}(P) ,$$

and from the relation

$$(T^* T e_0, e_i) = (e_0, T^* T e_i) = \lambda_i^2 (e_0, e_i) = 0 \quad \text{for } i \in I$$

it follows that $T^* T e_0$ also lies in $\mathfrak{R}(P)$. Therefore

$$T^* T e_0 = P T^* T P e_0 = T_0^* T_0 e_0 = \lambda_0^2 e_0 .$$

From this it follows that we could enlarge the maximal orthonormal system $[e_i, I]$ by adding the element $e_0$. But since this is impossible, the statement $\boldsymbol{\beta}(T_0) = 0$ or $T_0 = 0$ must be valid.

By using (*) we obtain the representation

$$T x = T \left[ \sum_I (x, e_i) e_i \right] + T P x = \sum_I (x, e_i) T e_i \quad \text{for } x \in E .$$

Now with $f_i = \lambda_i^{-1} T e_i$ we have

$$T x = \sum_I \lambda_i (x, e_i) f_i \quad \text{for } x \in E ,$$

and since

$$(f_i, f_j) = (\lambda_i^{-1} T e_i, \lambda_j^{-1} T e_j) = (\lambda_i^{-1} T^* T e_i, \lambda_j^{-1} e_j) = \lambda_i \lambda_j^{-1} \delta_{ij} ,$$

the elements $f_i$ form an orthonormal system in $F$.

Finally we show that the family of numbers $[\lambda_i, I]$ belongs to $\boldsymbol{c}_I$. For this purpose we consider the set

$$I_\delta = \{ i \in I : \lambda_i \geqq \delta \}$$

for an arbitrary positive number $\delta$. Then for two different indices $i, j \in I_\delta$ we always have

$$p_V(T e_i - T e_j)^2 = p_V(\lambda_i f_i - \lambda_j f_j)^2 = \lambda_i^2 + \lambda_j^2 \geqq 2\,\delta^2 .$$

However, since the set of all elements $T e_i$ with $i \in I$ is precompact, $I_\delta$ must be finite.

**Lemma.** *For each compact mapping $T \in \mathcal{L}(E, F)$ with $\lambda = \boldsymbol{\beta}(T) > 0$ there exists an element $e \in E$ with*

$$T^* T e = \lambda^2 e \quad \textit{and} \quad p_U(e) = 1 .$$

*Proof.* Since

$$\lambda = \sup \{p_V(T x) : x \in U\}$$

there is in $U$ a sequence of elements $x_n$ with

$$\lim_n p_V(T x_n) = \lambda .$$

Since the mapping $T$ is compact, we can assume that the sequence $\{T x_n\}$ converges to an element $f$ in $F$. If we set $T^* f = \lambda^2 e$, we have

$$\lim_n T^* T x_n = \lambda^2 e . \tag{1}$$

Since the right side of the identity

$$p_U(T^* T x_n - \lambda^2 x_n)^2 = p_U(T^* T x_n)^2 - 2\,\lambda^2\, p_V(T x_n)^2 + \lambda^4\, p_U(x_n)^2$$

converges to 0 as $n \to \infty$, we have

$$\lim_n (T^* T x_n - \lambda^2 x_n) = 0 . \tag{2}$$

By combining (1) and (2) we get

$$\lim_n x_n = e .$$

Consequently, we have the equation

$$T^* T e = \lim_n T^* T x_n = \lambda^2 e .$$

From the relation

$$\lambda = \lim_n p_V(T x_n) = p_V(T e) \leqq \lambda\, p_U(e) = \lambda \lim_n p_U(x_n) \leqq \lambda$$

we finally obtain the assertion $p_U(e) = 1$.

**8.3.2.** We now consider an arbitrary compact linear mapping $T$ from $E$ into $F$ and represent it in the way developed above:

$$T x = \sum \lambda_i(x, e_i)\, f_i \quad \text{for} \quad x \in E .$$

Since $[\lambda_i, I] \in \mathbf{c}_I$ and $\lambda_i > 0$ the sets

$$\mathfrak{i}_n = \left\{ i \in I : \lambda_i \geqq \frac{1}{n} \right\}$$

belong to $\mathfrak{F}(I)$ and we have

$$I = \bigcup_N \mathfrak{i}_n \, .$$

Consequently, the index set $I$ is at most countably infinite and we can assume that it has the form

$$I = \{0, 1, \ldots, n\} \quad \text{or} \quad I = \{0, 1, 2, \ldots\} \, .$$

We can also assume here that $i \leqq j$ always implies $\lambda_i \geqq \lambda_j$. Now if we set

$$\lambda_r(T) = \begin{cases} \lambda_r & \text{for } r \in I \\ 0 & \text{for } r \notin I \end{cases},$$

we get the

**Theorem.** *For each compact mapping* $T \in \mathcal{L}(E, F)$ *the identities*

$$\alpha_r(T) = \lambda_r(T)$$

*are valid.*

*Proof.* We write the mapping $T$ in the form

$$T x = \sum_I \lambda_i(T) \, (x, e_i) \, f_i \quad \text{for } x \in E$$

and for each number $r \in I$ define the mapping $A_0 \in \mathcal{A}_r(E, F)$ by the equation

$$A_0 \, x = \sum_{i=0}^{r-1} \lambda_i(T) \, (x, e_i) \, f_i \quad \text{for } x \in E \, .$$

Since

$$p_V(T x - A_0 \, x)^2 = \sum_{i \geqq r} \lambda_i(T)^2 \, |(x, e_i)|^2 \leqq \lambda_r(T)^2 \, p_U(x)^2 \quad \text{for } x \in E \, ,$$

we have

$$\alpha_r(T) \leqq \beta(T - A_0) \leqq \lambda_r(T) \, . \tag{1}$$

On the other hand, if $A$ is an arbitrary mapping from $\mathcal{A}_r(E, F)$, we can represent it in the form

$$A \, x = \sum_{k=1}^{r} (x, a_k) \, y_k \quad \text{for } x \in E$$

with elements $a_1, \ldots, a_r \in E$ and $y_1, \ldots, y_r \in F$. We now determine a non-trivial solution to the homogeneous system of equations

$$\sum_{i=0}^{r} \xi_i (e_i, a_k) = 0 \quad (k = 1, \ldots, r) \, ,$$

which we can normalize so that

$$\sum_{i=0}^{r} |\xi_i|^2 = 1$$

is true. For the element

$$x_0 = \sum_{i=0}^{r} \xi_i e_i$$

we then get the statement

$$A x_0 = \sum_{k=1}^{r} (x_0, a_k) y_k = \sum_{k=1}^{r} \sum_{i=0}^{r} \xi_i (e_i, a_k) y_k = o .$$

Finally since $p_U(x_0) = 1$ the estimate

$$\beta(T - A)^2 \geqq p_V(T x_0 - A x_0)^2 = p_V(T x_0)^2 = \sum_{i=0}^{r} \lambda_i(T)^2 |\xi_i|^2 \geqq \lambda_r(T)^2 .$$

holds. However, since $A$ was an arbitrary mapping from $\mathcal{A}_r(E, F)$, we obtain the inequality

$$\alpha_r(T) \geqq \lambda_r(T) .$$

By combining (1) and (2) we get the identity stated for all numbers $r \in I$.

If $r$ is not in $I$, then by definition we have $\lambda_r(T) = 0$. However, since the dimension of the range of $T$ is at most equal to $r$, $T$ belongs to $\mathcal{A}_r(E, F)$ and we also have $\alpha_r(T) = 0$.

**8.3.3. Theorem.** *If $E$ and $F$ are two Hilbert spaces then the nuclear mappings from $E$ into $F$ coincide with the mappings of type $\mathfrak{l}^1$, and for $T \in \mathcal{N}(E, F) = \mathfrak{l}^1(E, F)$ we have $\boldsymbol{\nu}(T) = \boldsymbol{\varrho}_1(T)$.*

*Proof.* We consider an arbitrary nuclear mapping $T \in \mathcal{N}(E, F)$ and by 8.3.1 represent it in the form

$$T x = \sum_I \lambda_i (x, e_i) f_i \quad \text{for } x \in E .$$

Since $T$ is nuclear, there are for each positive number $\delta$ elements $x_n \in E$ and $y_n \in F$ with

$$\sum_N p_U(x_n) p_V(y_n) \leqq \boldsymbol{\nu}(T) + \delta ,$$

and

$$T x = \sum_N (x, x_n) y_n \quad \text{for } x \in E .$$

Consequently, since

$$\lambda_i = (T e_i, f_i) = \sum_N (e_i, x_n) (y_n, f_i)$$

we get the estimate

$$\sum_I \lambda_i \leqq \sum_N \sum_I |(e_i, x_n) (y_n, f_i)| \leqq \sum_N \Big\{\sum_I |(e_i, x_n)|^2\Big\}^{1/2} \Big\{\sum_I |(y_n, f_i)|^2\Big\}^{1/2}$$
$$\leqq \sum_N p_U(x_n) p_V(y_n) \leqq \boldsymbol{\nu}(T) + \delta ,$$

from which we derive the inequality

$$\sum_I \lambda_i \leqq \boldsymbol{\nu}(T)$$

by taking the limit as $\delta \to 0$. On the other hand, since the index set $I$ is finite or countably infinite we also have

$$\boldsymbol{\nu}(T) \leqq \sum_I \lambda_i\, p_U(e_i)\, p_V(f_i) = \sum_I \lambda_i$$

by 3.1.1. Consequently,

$$\boldsymbol{\nu}(T) = \sum_I \lambda_i\,.$$

But since the relation

$$\boldsymbol{\varrho}_1(T) = \sum_R \alpha_r(T) = \sum_I \lambda_i$$

holds by Theorem 8.3.2, the identity

$$\boldsymbol{\varrho}_1(T) = \boldsymbol{\nu}(T)$$

is valid. Thus we have proved that each nuclear mapping is of type $\boldsymbol{l}^1$.

On the other hand, all mappings of type $\boldsymbol{l}^1$ must be nuclear because they can be represented as

$$T\,x = \sum \lambda_i(x, e_i)\, f_i \quad \text{for } x \in E$$

with

$$\sum_I \lambda_i\, p_U(e_i)\, p_V(f_i) < +\infty\,.$$

**8.3.4. Theorem.** *If $E$ and $F$ are two Hilbert spaces, then the Hilbert-Schmidt-mappings from $E$ into $F$ coincide with the mappings of type $\boldsymbol{l}^2$ and for $T \in \mathscr{S}(E, F) = \boldsymbol{l}^2(E, F)$ we have $\boldsymbol{\sigma}(T) = \boldsymbol{\varrho}_2(T)$.*

*Proof.* We consider an arbitrary compact mapping $T \in \mathscr{L}(E, F)$ and write it according to 8.3.1 as

$$T\,x = \sum \lambda_i(x, e_i)\, f_i \quad \text{for } x \in E\,.$$

If the orthonormal system $[e_i, I]$ is extended to a complete orthonormal system $[e_i, I']$ in $E$, we then have $T\,e_i = \bar{\bar{o}}$ for $i \in I' \backslash I$. Consequently, the identity

$$\boldsymbol{\sigma}(T)^2 = \sum_{I'} p_V(T\,e_i)^2 = \sum_I \lambda_i^2 = \sum_R \alpha_r(T)^2 = \boldsymbol{\varrho}_2(T)^2$$

holds by 8.3.2. Therefore, the two statements

$$\boldsymbol{\sigma}(T) < +\infty \quad \text{and} \quad \boldsymbol{\varrho}_2(T) < +\infty$$

are equivalent for each compact mapping $T \in \mathscr{L}(E, F)$. Our assertion is thus proved because Hilbert-Schmidt-mappings as well as mappings of type $\boldsymbol{l}^2$ are compact.

## 8.4. Nuclear and Absolutely Summing Mappings

**8.4.1.** For our further investigations we need the following

**Lemma 1.** *For each $r$-dimensional linear subspace $F$ of a normed space $E$ there are elements $x_1, \ldots, x_r \in F$ and linear forms $a_1, \ldots, a_r \in E'$ with $p_U(x_i) = 1$, $p_{U^0}(a_k) = 1$ and $\langle x_i, a_k\rangle = \delta_{ik}$. Then we have*

$$x = \sum_{i=1}^{r} \langle x, a_i\rangle\, x_i \quad \text{for all} \quad x \in F.$$

*Proof.* We consider an arbitrary system of linearly independent elements $y_1, \ldots, y_r$ in $F$ and set

$$\delta(b_1, \ldots, b_r) = |\det\{\langle y_i, b_k\rangle\}| \quad \text{for} \quad b_1, \ldots, b_r \in U^0 .$$

Then $\delta(b_1, \ldots, b_r)$ is a continuous function on the compact $r$-fold topological product of the weakly compact unit ball $U^0$ of $E'$. Consequently, there must exist elements $a_1, \ldots, a_r \in U^0$ for which $\delta(a_1, \ldots, a_r)$ assums the maximum $\delta_0$, which certainly must be greater than 0 since the elements $y_1, \ldots, y_r$ are linearly independent.

If the elements $x_1, \ldots, x_r \in F$ are the uniquely determined solution set of the system of equations

$$\sum_{j=1}^{r} \langle y_i, a_j\rangle\, x_j = y_i \qquad (i = 1, \ldots, r) ,$$

we then have

$$\langle x_j, a_k\rangle = \delta_{jk} .$$

Since

$$\sum_{j=1}^{r} \langle y_i, a_j\rangle \langle x_j, b_k\rangle = \langle y_i, b_k\rangle \qquad \text{for} \quad b_1, \ldots, b_r \in U^0 ,$$

we get

$$\delta(a_1, \ldots, a_r)\, |\det\{\langle x_j, b_k\rangle\}| = \delta(b_1, \ldots, b_r)$$

from the multiplication theorem for determinants. Therefore, the inequality

$$|\det\{\langle x_j, b_k\rangle\}| \leqq 1 \qquad \text{for} \quad b_1, \ldots, b_r \in U^0$$

holds. If we set $b_k = a_k$ for $k \neq i$ and $b_i = b \in U^0$ we then obtain

$$|\langle x_i, b\rangle| \leqq 1 \qquad \text{for } b \in U^0 \qquad \text{or} \qquad p_U(x_i) \leqq 1 .$$

Since $p_{U^0}(a_i) \leqq 1$ holds by hypothesis, it follows from

$$1 = \langle x_i, a_i\rangle \leqq p_U(x_i)\, p_{U^0}(a_i) ,$$

that

$$p_U(x_i) = 1 \qquad \text{and} \qquad p_{U^0}(a_i) = 1 .$$

Since the elements $x_1, \ldots, x_r$ form a linearly independent basis in $F$ we can write each element $x \in F$ uniquely as a linear combination

$$x = \sum_{i=1}^{r} \xi_i x_i .$$

Here we have

$$\langle x, a_k \rangle = \sum_{i=1}^{r} \xi_i \langle x_i, a_k \rangle = \xi_k .$$

Therefore, our assertion is proved.

For two arbitrary normed spaces $E$ and $F$ with closed unit balls $U$ and $V$ we now get the

**Lemma 2.** *Each mapping* $T \in \mathcal{A}(E, F)$ *with* $\dim \mathfrak{R}(T) = r$ *can be represented in the form*

$$T x = \sum_{i=1}^{r} \lambda_i \langle x, a_i \rangle y_i$$

*with linear forms* $a_i \in U^0$ *and elements* $y_i \in V$ *so that the inequality*

$$|\lambda_i| \leqq \beta(T)$$

*holds for the numbers* $\lambda_i$.

*Proof.* For the $r$-dimensional range of $T$ we determine elements $y_i \in \mathfrak{R}(T)$ and linear forms $b_i \in F'$ with the properties presented in Lemma 1. The identity

$$T x = \sum_{i=1}^{r} \langle x, T' b_i \rangle y_i \quad \text{for} \quad x \in E ,$$

then holds and we have

$$\lambda_i = p_{U^0}(T' b_i) > 0 .$$

Finally if we set

$$a_i = \lambda_i^{-1} T' b_i ,$$

we get the representation

$$T x = \sum_{i=1}^{r} \lambda_i \langle x, a_i \rangle y_i$$

for the mapping $T$. Moreover, we have

$$a_i \in U^0 , \quad y_i \in V \quad \text{and} \quad |\lambda_i| \leqq \beta(T) .$$

**8.4.2.** After these preparations we now obtain the

**Proposition.** *Each mapping* $T \in \mathfrak{l}^p(E, F)$ *with* $0 < p \leqq 1$ *can be represented as*

$$T x = \sum_R \lambda_r \langle x, a_r \rangle y_r$$

*with linear forms* $a_r \in U^0$ *and elements* $y_r \in V$, *such that the inequality*

$$\left\{\sum_R |\lambda_r|^p\right\}^{1/p} \leqq 2^{2+3/p}\, \varrho_p(T)$$

*holds for the numbers* $\lambda_r$.

*Proof*. For $n = 1, 2, \ldots$ we determine mappings $A_n \in \mathcal{A}_{2^n-2}(E, F)$ with

$$\beta(T - A_n) \leqq 2\,\alpha_{2^n-2}(T)$$

and set

$$B_n = A_{n+1} - A_n .$$

Then the statements

$$d_n = \dim \Re(B_n) \leqq 2^{n+2}$$

and

$$\beta(B_n) \leqq \beta(T - A_n) + \beta(T - A_{n+1}) \leqq 4\,\alpha_{2^n-2}(T)$$

are valid. Consequently, we have

$$d_n \beta(B_n)^p \leqq 2^{2p+n+2}\, \alpha_{2^n-2}(T)^p .$$

Since the sequence $[\alpha_r(T), R]$ decreases monotonically, the inequality

$$\sum_N 2^{n-1}\, \alpha_{2^n-2}(T)^p \leqq \sum_N \sum_{r=2^{n-1}-1}^{2^n-2} \alpha_r(T)^p = \sum_R \alpha_r(T)^p = \varrho_p(T)^p$$

holds. Therefore, the estimate

$$\sum_N d_n\, \beta(B_n)^p \leqq 2^{2p+3}\, \varrho_p(T)^p$$

is valid.

Using Lemma 2 from 8.4.1 we write the mappings $B_n$ in the form

$$B_n\, x = \sum_{i=1}^{d_n} \lambda_i^{[n]} \langle x, a_i^{[n]}\rangle\, y_i^{[n]} .$$

Here

$$a_i^{[n]} \in U^0 , \quad y_i^{[n]} \in V \quad \text{and} \quad |\lambda_i^{[n]}| \leqq \beta(B_n)$$

holds. Consequently, we have

$$\sum_N \sum_{i=1}^{d_n} |\lambda_i^{[n]}|^p \leqq \sum_N d_n\, \beta(B_n)^p \leqq 2^{2p+3}\, \varrho_p(T)^p .$$

Our assertion is thus proved because for all $x \in E$ the identity

$$T\,x = \lim_m A_{m+1}\, x = \sum_N B_n\, x = \sum_N \sum_{i=1}^{d_n} \lambda_i^{[n]} \langle x, a_i^{[n]}\rangle\, y_i^{[n]}$$

is true.

**8.4.3.** In 8.3.3 we have shown that in Hilbert spaces the nuclear mappings and mappings of type $\boldsymbol{l}^1$ are the same. In arbitrary normed spaces, however, there are nuclear mappings which are not of type $\boldsymbol{l}^1$. On the other hand, as an immediate consequence of Proposition 8.4.2 we get the

**Theorem.** *For $E$ and $F$ two normed spaces every mapping $T \in \boldsymbol{l}^1(E, F)$ is nuclear and we have $\boldsymbol{\nu}(T) \leqq 2^5\, \boldsymbol{\varrho}_1(T)$.*

**8.4.4. Problem.** *Determine the smallest positive number $\varrho$, for which the inequality*

$$\boldsymbol{\nu}(T) \leqq \varrho\, \boldsymbol{\varrho}_1(T)$$

*holds for all mappings $T$ of type $\boldsymbol{l}^1$.*

**8.4.5.** As a converse to the statement of Theorem 8.4.3 we obtain for three arbitrary normed spaces $E$, $F$ and $G$ the following

**Theorem.** *The product $S\,T$ of two absolutely summing mappings $T \in \mathcal{P}(E, F)$ and $S \in \mathcal{P}(F, G)$ is of type $\boldsymbol{l}^2$ and we have*

$$\boldsymbol{\varrho}_2(S\,T) \leqq \boldsymbol{\pi}(S)\, \boldsymbol{\pi}(T)\,.$$

*Proof.* As in 3.3.5 decompose $S\,T$ considered as a mapping from $E$ into $\tilde{G}$ into a product

$$S\,T = S_2\, K_S\, \tilde{S}_1\, T_2\, K_T\, T_1\,.$$

Then because of Theorem 8.3.4 the inequality

$$\boldsymbol{\varrho}_2(K_S\, \tilde{S}_1\, T_2) = \boldsymbol{\sigma}(K_S\, \tilde{S}_1\, T_2) \leqq \boldsymbol{\pi}(T)$$

is valid. Therefore, we have

$$\boldsymbol{\varrho}_2^{\tilde{G}}(S\,T) \leqq \boldsymbol{\beta}(S_2)\, \boldsymbol{\varrho}_2(K_S\, \tilde{S}_1\, T_2)\, \boldsymbol{\beta}(K_T\, T_1) \leqq \boldsymbol{\pi}(S)\, \boldsymbol{\pi}(T)\,.$$

Our assertion is thus proved because $S\,T$ is also, as a mapping from $E$ into $G$, of type $\boldsymbol{l}^2$ on the basis of Proposition 8.1.3, and $\boldsymbol{\varrho}_2^G(S\,T) = \boldsymbol{\varrho}_2^{\tilde{G}}(S\,T)$.

**8.4.6.** In the same way as in 8.4.5 we can show that the product of three absolutely summing mappings is of type $\boldsymbol{l}^1$.

## 8.5. Mappings of Type s

**8.5.1.** For two normed spaces $E$ and $F$ we consider the collection $\mathbf{s}(E, F)$ of all mappings $T \in \mathcal{L}(E, F)$ for which the inequality

$$\sum_R \alpha_r(T)^p < +\infty$$

holds for every positive number $p$, and we say that these mappings are of **type s**.

**8.5.2.** From the relation

$$\mathbf{s}(E, F) = \bigcap_{p>0} \mathbf{l}^p(E, F)$$

we get the

**Proposition.** *$\mathbf{s}(E, F)$ is a linear space.*

**8.5.3.** We obtain a metric topology on $\mathbf{s}(E, F)$ by using the sets

$$U_{p,\varepsilon}(T) = \{S \in \mathbf{s}(E, F) : \varrho_p(S - T) \leqq \varepsilon\} \quad \text{with } p > 0 \text{ and } \varepsilon > 0$$

as a fundamental system of neighborhoods of the mapping $T$.

The following two propositions are immediate consequences of 8.2.4 and 8.2.5.

**Proposition.** *For a normed space $E$ and a Banach space $F$, $\mathbf{s}(E, F)$ is complete.*

**Proposition.** *The linear space $\mathcal{A}(E, F)$ is dense in $\mathbf{s}(E, F)$.*

**8.5.4.** For $E$, $F$ and $G$ three normed spaces, we have the

**Proposition.** (1) *$T \in \mathcal{L}(E, F)$ and $S \in \mathbf{s}(F, G)$ implies $S\,T \in \mathbf{s}(E, G)$.*
(2) *$T \in \mathbf{s}(E, F)$ and $S \in \mathcal{L}(F, G)$ implies $S\,T \in \mathbf{s}(E, G)$.*

**8.5.5.** A sequence of numbers $[\lambda_r, R]$ is called **rapidly decreasing** if the sequences of numbers $[(r + 1)^k \lambda_r, R]$ is bounded for $k = 0, 1, \ldots$.

**Lemma 1.** *If $[\lambda_r, R]$ is a rapidly decreasing sequence of numbers the inequality*

$$\sum_R (r + 1)^k |\lambda_r|^p < +\infty$$

*holds for all numbers $k = 0, 1, \ldots$ and $p > 0$.*

*Proof.* We determine a natural number $h$ with $h\,p \geqq k + 2$. Since there is, by hypothesis, a positive number $\varrho$ with

$$(r + 1)^h |\lambda_r| \leqq \varrho \quad \text{for } r \in R$$

we have

$$\sum_R (r + 1)^k |\lambda_r|^p \leqq \varrho^p \sum_R (r + 1)^{-2} < +\infty\,.$$

**Lemma 2.** *Every sequence of numbers $[\lambda_r, R]$ with $|\lambda_0| \geqq |\lambda_1| \geqq \cdots \geqq 0$, for which*

$$\sum_R |\lambda_r|^p < +\infty$$

*holds for all positive numbers $p$, is rapidly decreasing.*

*Proof.* Our assertion follows immediately from the inequality

$$(s + 1)\,|\lambda_s|^p \leqq \sum_{r=0}^{s} |\lambda_s|^p \leqq \sum_R |\lambda_s|^p < +\infty$$

which is valid for $s = 0, 1, \ldots$

From these preliminaries we obtain the

**Proposition.** *A mapping $T \in \mathfrak{L}(E, F)$ is of type* **s** *if and only if the sequence of its approximation numbers is rapidly decreasing.*

**8.5.6.** For two normed spaces with closed unit balls $U$ and $V$ we have the

**Theorem.** *A mapping $T \in \mathfrak{L}(E, F)$ is of type* **s** *if and only if it can be represented in the form*

$$T x = \sum_R \lambda_r \langle x, a_r \rangle y_r$$

*with linear forms $a_r \in U^0$, elements $y_r \in V$ and a rapidly decreasing sequence of numbers $[\lambda_r, R]$.*

*Necessity.* If $T$ is a mapping in $\mathbf{s}(E, F)$, we determine mappings $A_s \in \mathfrak{A}_s(E, F)$ with

$$\beta(T - A_s) \leqq 2\,\alpha_s(T) \qquad \text{for} \quad s \in S = \{0, 1, \ldots\}$$

Then the statements

$$d_s = \dim \Re(B_s) \leqq 2\,s + 1 \qquad \text{and} \qquad \beta(B_s) \leqq 4\,\alpha_s(T)$$

are valid for the mappings $B_s = A_{s+1} - A_s$, and we have

$$\sum_S d_s\, \beta(B_s) \leqq 2 \cdot 4^p \sum_S (s+1)\, \alpha_s(T)^p < +\infty \qquad \text{for all } p > 0\,.$$

Using Lemma 2 from 8.4.1 we put the mapping $B_s$ into the form

$$B_s x = \sum_{i=1}^{d_s} \lambda_i^{[s]} \langle x, a_i^{[s]} \rangle y_i^{[s]}\,,$$

such that

$$a_i^{[s]} \in U^0\,, \qquad y_i^{[s]} \in V \qquad \text{and} \qquad |\lambda_i^{[s]}| \leqq \beta(B_s)\,.$$

Consequently we have

$$\sum_S \sum_{i=1}^{d_s} |\lambda_i^{[s]}|^p \leqq \sum_S d_s\, \beta(B_s)^p < +\infty \qquad \text{for all } p > 0\,,$$

and for each element $x \in E$ the identity

$$T x = \lim_r A_{r+1} x = \sum_S B_s x = \sum_S \sum_{i=1}^{d_s} \lambda_i^{[s]} \langle x, a_i^{[s]} \rangle y_i^{[s]}$$

holds. We have thus shown that the mapping $T$ can be represented as

$$T x = \sum_R \lambda_r \langle x, a_r \rangle y_r$$

with linear forms $a_r \in U^0$ and elements $y_r \in V$ such that for each positive number $p$ the inequality

$$\sum_R |\lambda_r|^p < +\infty$$

holds. Since we can always reorder $R$ so that $|\lambda_0| \geqq |\lambda_1| \geqq \ldots \geqq 0$, the sequence $[\lambda_r, R]$ is rapidly decreasing.

*Sufficiency.* We now consider a mapping $T \in \mathcal{L}(E, F)$ which can be represented in the given way. Since the mapping $A$ with

$$A\,x = \sum_{r=0}^{s-1} \lambda_r \langle x, a_r \rangle\, y_r \quad \text{for} \quad x \in E$$

belongs to $\mathcal{A}_s(E, F)$, we have

$$\alpha_s(T) \leqq \beta(T - A) \leqq \sum_{r=s}^{\infty} |\lambda_r| \,.$$

Consequently, for all $p$ between 0 and 1, the inequality

$$\alpha_s(T)^p \leqq \left(\sum_{r=s}^{\infty} |\lambda_r|\right)^p \leqq \sum_{r=s}^{\infty} |\lambda_r|^p$$

is valid, and we get the estimate

$$\sum_S \alpha_s(T)^p \leqq \sum_S \sum_{r=s}^{\infty} |\lambda_r|^p = \sum_R (r + 1)\, |\lambda_r|^p < + \infty \,.$$

Therefore $T$ is of type **s**.

## 8.6. A Characterization of Nuclear Locally Convex Spaces

**8.6.1. Theorem.** *A locally convex space $E$ is nuclear if and only if for some, resp. each, positive number $p$ the following statement is valid:*

*For each zero neighborhood $U \in \mathfrak{U}(E)$ there is a zero neighborhood $V \in \mathfrak{U}(E)$ with $V < U$ such that the canonical mapping from $E(V)$ onto $E(U)$ is of type $\boldsymbol{l}^p$.*

*Necessity.* For an arbitrary zero neighborhood $U = U_0 \in \mathfrak{U}(E)$ we determine zero neighborhoods $U_1, \ldots, U_{4n} \in \mathfrak{U}(E)$ with $U_{4n} < \cdots < U_1 < U_0$ such that the canonical mappings $E(U_k, U_{k-1})$ are absolutely summing, and set $V = U_{4n}$. Then the mapping

$$E(V, U) = E(U_1, U_0) \cdots E(U_{4n}, U_{4n-1})$$

is of type $\boldsymbol{l}^{1/n}$ by Theorem 8.2.7, because we can combine each pair of consecutive mappings in the product on the right-hand side to obtain a mapping of type $\boldsymbol{l}^2$.

If $p$ is an arbitrary positive number, we choose $n$ greater than $1/p$. Then the canonical mapping $E(V, U)$ is of type $\boldsymbol{l}^p$.

*Sufficiency.* If the stated assertion is satisfied for some positive number $p$, we determine for a zero neighborhood $U = U_0 \in \mathfrak{U}(E)$, zero neighborhoods $U_1, \ldots, U_n \in \mathfrak{U}(E)$ with $U_n < \cdots U_1 < U_0$, such that the canonical mappings $E(U_k, U_{k-1})$ are of type $\boldsymbol{l}^p$. Here, the natural number $n$ is assumed to be greater than $p$. We now set $V = U_n$. Then the mapping

$$E(V, U) = E(U_1, U_0) \cdots E(U_n, U_{n-1})$$

is of type $\mathbf{l}^{p/n}$ by Theorem 8.2.7. But since $p/n < 1$, $E(V, U)$ must be nuclear by Theorem 8.4.3.

**8.6.2.** For dual metric locally convex spaces we also have the following

**Theorem.** *A dual metric locally convex space $E$ is nucelar if and only if for each zero neighborhood $U \in \mathfrak{U}(E)$ there is a zero neighborhood $V \in \mathfrak{U}(E)$ with $V < U$ such that the canonical mapping from $E(V)$ onto $E(U)$ is of type* $\mathbf{s}$.

*Proof.* Since the sufficiency of the given assertion follows immediately from Theorem 8.6.1 we have only to prove its necessity. For this purpose, we determine a sequence of zero neighborhoods $V_n \in \mathfrak{U}(E)$ with $V_n < U$ such that the canonical mapping from $E(V_n)$ onto $E(U)$ is of type $\mathbf{l}^{1/n}$. We now consider in $E$ a fundamental system of bounded subsets $B_n$ with $B_1 \subset B_2 \subset \cdots$. Then there exist positive numbers $\varrho_{mn}$ with $B_m \subset \varrho_{mn} V_n$, and we can form the closed and absolutely convex set

$$V = \bigcap_N \varrho_{nn} V_n .$$

Since the statement

$$B_m \subset \sup \{\varrho_{mn} \varrho_{nn}^{-1} : n = 1, \ldots, m\} V \quad \text{for } m = 1, 2, \ldots$$

is valid, there exists for each bounded subset $B$ of $E$ a positive number $\varrho$ with $B \subset \varrho V$. But since the nuclear dual metric space $E$ is quasi-barrelled by Lemma 4.4.12, $V$ must be a zero neighborhood in $E$. The canonical mapping from $E(V)$ onto $E(U)$ is of type $\mathbf{s}$ because it can be represented in the form

$$E(V, U) = E(V_n, U)\, E(V, V_n)$$

with $n = 1, 2, \ldots$.

**8.6.3** The sequence space $\Sigma$ is easily seen to be an example of a nuclear metric space in which there are zero neighborhoods $U \in \mathfrak{U}(E)$ with no zero neighborhood $V \in \mathfrak{U}(E)$ such that $V < U$ and the canonical mapping from $E(V)$ onto $E(U)$ is of type $\mathbf{s}$.

**8.6.4.** By applying the same reasoning as in 8.6.1 we get the

**Theorem.** *A locally convex space $E$ is dual nuclear if and only if for some resp. each positive number $p$ the following statement is valid:*

*For each set $A \in \mathfrak{B}(E)$ there is a set $B \in \mathfrak{B}(E)$ with $A < B$ such that the canonical mapping from $E(A)$ into $E(B)$ is of type* $\mathbf{l}^p$.

**8.6.5.** As the dual analog to the assertion of 8.6.2 we get the

**Theorem.** *A metric locally convex space $E$ is dual nuclear if and only if for each set $A \in \mathfrak{B}(E)$ there is a set $B \in \mathfrak{B}(E)$ with $A < B$ such that the canonical mapping from $E(A)$ into $E(B)$ is of type* $\mathbf{s}$.

**8.6.6.** As a substantial improvement of Proposition 4.4.3 we derive the following

**Proposition.** *For a nuclear or dual nuclear locally convex space $E$, all canonical mappings $E(A, U)$ with $A \in \mathfrak{B}(E)$ and $U \in \mathfrak{U}(E)$ are of type* $\mathbf{s}$.

*Proof.* If $V$ is an arbitrary zero neighborhood from $\mathfrak{U}(E)$ with $V < U$ we have

$$E(A, U) = E(V, U)\, E(A, V)\,.$$

On the basis of Theorem 8.6.1 we can, in the case of a nuclear space, choose $V$ in such a way that $E(V, U)$ is of type $\boldsymbol{l}^p$ for an arbitrary positive number $p$. Consequently, the canonical mapping $E(A, U)$ must be of type $\mathbf{s}$.

The proof of our assertion for dual nuclear locally convex spaces proceeds in the same way if we write the canonical mapping $E(A, U)$ in the form

$$E(A, U) = E(B, U)\, E(A, B)$$

with a set $B \in \mathfrak{B}(E)$.

**8.6.7.** As a supplement to 4.4.6 we obtain finally the

**Theorem.** *A metric or dual metric locally convex space $E$ is nuclear if and only if every canonical mapping $E(K, U)$ with $K \in \mathfrak{K}(E)$ and $U \in \mathfrak{U}(E)$ is of type* $\mathbf{s}$.

Chapter 9

# Diametral and Approximative Dimension

In this chapter invariants are constructed for locally convex spaces. We treat the *approximative dimension* of A. N. Kolmogorov [2] and A. Pełczyński [1] and the closely related *diametral dimension* (C. Bessaga, A. Pełczyński and S. Rolewicz [1], [2]), which is especially suited to investigation of locally convex sequence spaces.

As the suggestion of I. M. Gelfand, B. S. Mitiagin [2], [4] characterized nuclear locally convex spaces with the help of these dimensions (Theorem 9.4.1 and 9.8.4). Special cases had already been investigated by S. Rolewicz [1] as well as A. S. Dynin and B. S. Mitiagin [1].

Two fundamental concepts for bounded subsets in normed spaces are introduced in 9.1.1 and 9.6.1. Here we treat the *diameters* of A. N. Kolmogorov and the so-called *ε-content,* whose definition goes back to L. Pontrjagin and L. Schnirelmann. The connection between the two concepts, established in 9.6.3, was discovered by B. S. Mitiagin [2], [4].

In 9.5.3 we use the method developed to prove that each bounded subset of a nuclear $(F)$-space is contained in the closed and absolutely convex hull of a rapidly decreasing sequence of elements. This theorem is due to A. Grothendieck who had already noted that "*propriétés de décroissance*" of equicontinuous subsets of the topological dual could be used to classify nuclear convex spaces ([3], Chap. II, p. 64).

One might try to construct a theory of nuclear locally convex spaces using only the concept of approximative dimension. For such a construction it is essential that the properties of approximative dimension given in 9.7.5 carry over to the strong topological dual. However, except for some relations stated without proof by C. Bessaga, A. Pełczyńksi and S. Rolewicz [1], little has been done in this direction. This large gap ought to be filled as soon as possible (Problem 9.7.6).

## 9.1. The Diameters of Bounded Subsets in Normed Spaces

**9.1.1.** Let $B$ be an arbitrary bounded subset in a normed space $E$ with closed unit ball $U$. The infimum $\delta_r(B)$ of all positive numbers $\delta$ such that there is a linear subspace $F$ of $E$ with dimension at most $r$ for which

$$B \subset \delta U + F$$

is designated as the **r-th diameter** of $B$. We have

$$\delta_0(B) \geqq \delta_1(B) \geqq \delta_2(B) \geqq \cdots \geqq 0 .$$

**9.1.2. Lemma.** *If $P \in \mathcal{L}(E, F)$ is a mapping with $\beta(P) \leqq 1$, $P^2 = P$ and* dim $\mathfrak{R}(P) = r + 1$, *then for each bounded subset $B$ of $E$,*

$$U \cap \mathfrak{R}(P) \subset \delta^{-1} B$$

*implies*

$$\delta_r(B) \geqq \delta .$$

*Proof.* We assume that $\delta_r(B) < \delta$ and choose a number $\delta_0$ lying between $\delta_r(B)$ and $\delta$. Then there exists a linear subspace $F$ of $E$ of at most $r$-dimension with

$$B \subset \delta_0 U + F .$$

Since $F_0 = P(F)$ is a proper linear subspace of $\mathfrak{R}(P)$, there exists an element $x_0 \in \mathfrak{R}(P)$ with

$$\varrho = \inf \{p_U(x_0 - y) : y \in F_0\} > 0 .$$

We now choose an element $y_0 \in F_0$ with

$$\varrho_0 = p_U(x_0 - y_0) < \varrho\, \delta_0^{-1}\, \delta .$$

Since $P(U) \subset U$, the relation

$$U \cap \mathfrak{R}(P) \subset \delta^{-1} P(B) \subset \delta^{-1} \delta_0 U + F_0$$

holds implying that the element $\varrho_0^{-1}(x_0 - y_0)$ of $U \cap \mathfrak{R}(P)$ can be represented as

$$\varrho_0^{-1}(x_0 - y_0) = \delta^{-1} \delta_0 x + y \quad \text{with} \quad x \in U \quad \text{and} \quad y \in F_0 .$$

Consequently, we have

$$p_U(x_0 - y_0 - \varrho_0 y) = \varrho_0 \delta^{-1} \delta_0 p_U(x) < \varrho .$$

However, from the definition of $\varrho$ and the fact that $y_0 + \varrho_0 y \in F_0$, we find

$$p_U(x_0 - y_0 - \varrho_0 y) \geqq \varrho .$$

This contradiction shows that the relation $\delta_r(B) < \delta$ is false.

**9.1.3. Proposition.** *For the bounded subset*

$$B = \left\{[\xi_r, R] : \sum_R \delta_r^{-1} |\xi_r| \leqq 1\right\}$$

*of $l_R^1$ determined by the numbers $\delta_0 \geqq \delta_1 \geqq \delta_2 \geqq \cdots > 0$ we have the identities*

$$\delta_r(B) = \delta_r .$$

*Proof.* Consider the linear subspace

$$F_s = \{[\xi_r, R] : \xi_r = 0 \quad \text{for } r \geqq s\}$$

of $\boldsymbol{l}_R^1$. The relations

$$B \subset \delta_s U + F_s$$

hold with $U$ the closed unit ball of $\boldsymbol{l}_R^1$. Consequently,

$$\delta_s(B) \leqq \delta_s .$$

We now set $\mathfrak{r}_s = \{0, 1, \ldots, s\}$. Then the mappings $P_s$ defined by

$$P_s[\xi_r, R] = [\xi_r(\mathfrak{r}_s), R]$$

satisfy the hypotheses of the previous lemma, and we have $F_{s+1} = \mathfrak{N}(P_s)$. Moreover, since

$$U \cap F_{s+1} \subset \delta_s^{-1} B ,$$

we have

$$\delta_s(B) \geqq \delta_s .$$

**9.1.4. Proposition.** *A bounded subset $B$ of a normed space $E$ is precompact if and only if*

$$\lim_r \delta_r(B) = 0 .$$

*Proof.* We will first assume that $B$ is precompact. Then for each positive number $\delta$ there are finitely many elements $x_1, \ldots, x_s \in E$ with

$$B \subset \bigcup_{n=1}^{s} \{x_n + \delta U\} .$$

If $F$ is the linear subspace of $E$ with $\dim F \leqq s$ determined by these elements, we have

$$B \subset \delta U + F .$$

Consequently,

$$\delta_r(B) \leqq \delta \quad \text{for } r \geqq s .$$

Since $\delta$ was arbitrary,

$$\lim_r \delta_r(B) = 0 .$$

Conversely, consider a bounded subset $B$ with

$$\lim_r \delta_r(B) = 0 .$$

Then for each positive number $\delta$ there is a natural number $r$ with $\delta_r(B) < \delta$. By the definition of the $r$-th diameter there must exist a linear subspace $F$ of $E$ with dimension at most $r$ and

$$B \subset \delta U + F .$$

Since $U_0 = F \cap U$ as a bounded subset of a finite dimensional locally convex space is precompact, there are finitely many elements $x_1, \ldots x_s$ with

$$[\delta + \delta_0(B)]\, U_0 \subset \bigcup_{n=1}^{s} \{x_n + \delta U\} .$$

We now consider an arbitrary element $x \in B$ and represent it in the form

$$x = \delta y + z \quad \text{with} \quad y \in U \quad \text{and} \quad z \in F .$$

We then get the estimate

$$p_U(z) \leqq \delta p_U(y) + p_U(x) \leqq \delta + \delta_0(B)$$

for the norm of $z$. Consequently, we have

$$x = \delta y + z \in \delta U + [\delta + \delta_0(B)]\, U_0 \subset \bigcup_{n=1}^{s} \{x_n + 2\,\delta U\} .$$

We have thus shown that for each positive number $\delta$ there are finitely many elements $x_1, \ldots, x_s \in E$ with

$$B \subset \bigcup_{n=1}^{s} \{x_n + 2\,\delta U\} .$$

Therefore, the bounded subset $B$ is precompact.

**9.1.5. Proposition.** *A bounded subset $B$ of a normed space $E$ lies in a linear subspace of dimension at most $r$ if and only if*

$$\delta_r(B) = 0 .$$

*Proof.* If the bounded subset $B$ is contained in a linear subspace $F$ of $E$ of dimension at most $r$, then the relation

$$B \subset \delta U + F$$

holds for all positive numbers $\delta$. Consequently,

$$\delta_r(B) = 0 .$$

We now consider a bounded subset $B$ with $\delta_r(B) = 0$ and assume that there are at least $r + 1$ linearly independent elements $x_1, \ldots, x_{r+1} \in B$. By the Hahn Banach Theorem 0.4.4, we can determine linear forms $a_1, \ldots, a_{r+1} \in E'$ with $\langle x_i, a_k \rangle = \delta_{ik}$.

Since $\det \{\delta_{ik}\} = 1$, there exists a positive number $\sigma$ with

$$\det \{\delta_{ik} - \alpha_{ik}\} \neq 0 \quad \text{for} \quad |\alpha_{ik}| \leqq \sigma .$$

Set

$$\delta = \sigma \max \{p_{U^0}(a_k) : k = 1, \ldots, r + 1\}^{-1} .$$

Since $\delta_r(B) = 0$, there is a linear subspace $F$ of $E$ of dimension at most $r$ with

$$B \subset \delta U + F .$$

Consequently, we can represent each element $x_i \in B$ as

$$x_i = \delta y_i + z_i \quad \text{with} \quad y_i \in U \quad \text{and} \quad z_i \in F .$$

Since the elements $z_1, \ldots, z_{r+1}$ are linearly dependent, we have

$$\det \{\langle z_i, a_k \rangle\} = 0 .$$

On the other hand, since

$$\delta |\langle y_i, a_k \rangle| \leqq \delta p_{U^0}(a_k) \leqq \sigma ,$$

the relation

$$\det \{\langle z_i, a_k \rangle\} = \det \{\langle x_i, a_k \rangle - \delta \langle y_i, a_k \rangle\} \neq 0$$

must hold.

From the contradiction thus obtained we conclude that in $B$ there can be at most $r$ linearly independent elements. Our assertion is thus proved.

**9.1.6.** Let $E$ and $F$ are two normed spaces with closed unit balls $U$ and $V$. We consider an arbitrary mapping $T \in \mathfrak{L}(E, F)$ and investigate the relation between the approximation numbers $\alpha_r(T)$ defined in 8.1.1 and the diameters of the bounded subset

$$T(U) = \{T x : x \in U\}$$

of $F$. We obtain the following

**Lemma.** *For each mapping* $T \in \mathfrak{L}(E, F)$

$$\delta_r(T(U)) \leqq \alpha_r(T) \leqq (r + 1)\, \delta_r(T(U)) .$$

*Proof.* For a given arbitrary positive number $\varepsilon$ we determine a mapping $A \in \mathfrak{A}_r(E, F)$ with

$$\beta(T - A) \leqq \alpha_r(T) + \varepsilon .$$

From the relation

$$T(U) \subset \beta(T - A)\, V + \mathfrak{R}(A)$$

we get the estimate

$$\delta_r(T(U)) \leqq \beta(T - A) \leqq \alpha_r(T) + \varepsilon .$$

But since the positive number $\varepsilon$ can be taken arbitrarily small, we also have

$$\delta_r(T(U)) \leqq \alpha_r(T) .$$

In order to prove the second inequality we determine for the number $\delta = \delta_r(T(U)) + \varepsilon$ with $\varepsilon > 0$ a linear subspace $G$ of $F$ of dimension at most $r$ for which

$$T(U) \subset \delta V + G$$

holds. Then by Lemma 8.4.1 there are elements $z_1, \ldots, z_s \in G$ and linear forms $b_1, \ldots, b_s \in F'$ with $p_V(z_i) = 1$, $p_{V^0}(b_k) = 1$ and $\langle z_i, b_k\rangle = \delta_{ik}$ such that the identity

$$z = \sum_{i=1}^{s} \langle z, b_i\rangle z_i$$

holds. By means of

$$P y = \sum_{i=1}^{s} \langle y, b_i\rangle z_i \quad \text{for } y \in F$$

we define a linear mapping $P \in \mathcal{A}_r(E, F)$ with $\beta(P) \leqq s \leqq r$. We see that the mapping $A = P\,T$ belongs to $\mathcal{A}_r(E, F)$.

If $x$ is an arbitrary element of $U$ then $T\,x$ can be represented as

$$T\,x = \delta y + z \quad \text{with} \quad y \in V \quad \text{and} \quad z \in G\,.$$

Since $P\,z = z$, we get the estimate

$$p_V(T\,x - A\,x) = p_V(\delta y + z - \delta P\,y - P\,z) = \delta p_V(y - P y) \leqq \delta(r + 1)\,.$$

Hence we have

$$\alpha_r(T) \leqq \beta(T - A) \leqq (r + 1)\,\big(\delta_r(T(U)) + \varepsilon\big)\,,$$

and by taking the limit as $\varepsilon \to 0$ we get the stated inequality.

## 9.2. The Diametral Dimension of Locally Convex Spaces

**9.2.1.** In an arbitrary locally convex space $E$ we consider two zero neighborhoods $U, V \in \mathfrak{U}(E)$ with $V < U$. The infimum $\delta_r(V, U)$ of all positive numbers $\delta$, for which there is a linear subspace $F$ of $E$ of dimension at most $r$ with

$$V \subset \delta U + F$$

is designated as the **$r$-th diameter of $V$ with respect to $U$**. It can be shown that $\delta_r(V, U)$ coincides with the $r$-th diameter of the bounded subset

$$V(U) = \{x(U) : x \in V\}$$

of the normed space $E(U)$.

**9.2.2.** The **diametral dimension** $\Delta_{\mathfrak{U}}(E)$ of a locally convex space $E$ is the collection of all sequences $[\delta_r, R]$ of non negative numbers with the property that for each zero neighborhood $U \in \mathfrak{U}(E)$ there is a zero neighborhood $V \in \mathfrak{U}(E)$ with $V < U$ and

$$\delta_r(V, U) \leqq \delta_r \quad \text{for } r \in R\,.$$

It is easy to show that we can replace $\mathfrak{U}(E)$ with an arbitrary fundamental system of zero neighborhood in this definition.

**9.2.3.** Two locally convex spaces $E$ and $F$ are designated as **isomorphic** if there exists a one-to-one linear mapping from $E$ onto $F$ which is continuous in both directions. Since

$$\mathfrak{U}(F) = \{T(U) : U \in \mathfrak{U}(E)\}$$

and

$$\delta_r(T(V), T(U)) = \delta_r(V, U) \quad \text{for } V, U \in \mathfrak{U}(E) \quad \text{with } V < U,$$

we get the

**Theorem.** *For two isomorphic locally convex spaces $E$ and $F$*

$$\Delta_{\mathfrak{U}}(E) = \Delta_{\mathfrak{U}}(F).$$

**9.2.4.** The diametral dimension of a locally convex space is a meaningful generalization of the ordinary dimension as is shown by the

**Theorem.** *A locally convex space $E$ is at most $s$-dimensional if and only if the following assertion is valid:*

*There is in $\Delta_{\mathfrak{U}}(E)$ a sequence $[\delta_r^0, R]$ with $\delta_s^0 = 0$.*

*Proof.* If $E$ is a locally convex space of dimension at most $s$, then the relation $\delta_s(U, U) = 0$ is valid for each zero neighborhood $U \in \mathfrak{U}(E)$. Consequently, $\Delta_{\mathfrak{U}}(E)$ contains all sequences $[\delta_r, R]$ with $\delta_r > 0$ for $r < s$ and $\delta_r = 0$ for $r \geqq s$.

On the other hand, if the dimension of the locally convex space $E$ is greater than $s$, there exist $s + 1$ linearly independent elements $x_1, \ldots, x_{s+1} \in E$; and it is clear that the closed set

$$A = \left\{ \sum_{i=1}^{s+1} \xi_i x_i : \sum_{i=1}^{s+1} |\xi_i| = 1 \right\}$$

does not contain zero. There is thus a zero neighborhood $U \in \mathfrak{U}(E)$ with $A \cap U = \emptyset$. But then the elements $x_1(U), \ldots, x_{s+1}(U)$ are linearly independent in $E(U)$ because if the identity

$$\sum_{i=1}^{s+1} \xi_i x_i(U) = o(U) \quad \text{with } \sum_{i=1}^{s+1} |\xi_i| = 1$$

held it would result in the false assertion

$$\sum_{i=1}^{s+1} \xi_i x_i \in N(U) \subset U.$$

If $V$ is an arbitrary zero neighborhood in $\mathfrak{U}(E)$ with $V < U$, there is a positive number $\varrho$ with

$$x_i(U) \in \varrho V(U) \quad \text{for } i = 1, \ldots, s + 1.$$

Thus we have shown that the bounded subset $V(U)$ of the normed space $E(U)$ contains at least $s + 1$ linearly independent elements. Consequently, the inequality

$$\delta_s(V, U) = \delta_s(V(U)) > 0 \quad \text{for all } V \in \mathfrak{U}(E) \quad \text{with } V < U,$$

holds by Proposition 9.1.5 and the diametral dimension $\varDelta_{\mathfrak{U}}(E)$ can contain no sequence $[\delta_r^0, R]$ with $\delta_s^0 = 0$.

**9.2.5.** A reasonable dimension should reflect the size of the space considered. Unfortunately, it has not yet been proved without additional assumptions that

$$\varDelta_{\mathfrak{U}}(F) \subset \varDelta_{\mathfrak{U}}(E)$$

holds for each linear subspace $F$ of a locally convex space $E$. However, we do know that the statement

$$\varDelta_{\mathfrak{U}}(E/F) \subset \varDelta_{\mathfrak{U}}(E)$$

is valid for each quotient space $E/F$.

**9.2.6. Problem.** *Under what hypotheses on the locally convex space E does the identity*

$$\varDelta_{\mathfrak{U}}(E'_b) = \varDelta_{\mathfrak{U}}(E)$$

*hold?*

**9.2.7.** The **dual diametral dimension** $\varDelta_{\mathfrak{B}}(E)$ of a locally convex space $E$ is the collection of all sequences $[\delta_r, R]$ of non negative numbers which have the property that for each set $A \in \mathfrak{B}(E)$ there is a set $B \in \mathfrak{B}(E)$ with $A < B$ and

$$\delta_r(A, B) \leqq \delta_r \quad \text{for } r \in R\,.$$

Here the $r$-th diameter $\delta_r(A, B)$ of $A$ with respect to $B$ is defined exactly as in 9.2.1.

It is not yet known whether the dual diametral dimension always coincides with the diametral dimension of the strong topological dual.

**Problem.** *Under what hypotheses on the locally convex space E does the identity*

$$\varDelta_{\mathfrak{B}}(E) = \varDelta_{\mathfrak{U}}(E'_b)$$

*hold?*

## 9.3. The Diametral Dimension of Power Series Spaces

**9.3.1.** In each power series space $\varLambda$, determined by the system of all sequences $[\varrho^{\alpha_r}, R]$ with $0 < \varrho < \varrho_0$, the sets

$$\varepsilon\, U_\varrho = \left\{[\xi_r, R] \in \varLambda : \sum_R |\xi_r|\, \varrho^{\alpha_r} \leqq \varepsilon\right\} \quad \text{with } 0 < \varrho < \varrho_0 \text{ and } \varepsilon > 0\,.$$

form a fundamental system of zero neighborhoods. Since we can obtain

$$\delta_r(U_\sigma, U_\varrho) = (\sigma^{-1}\varrho)^{\alpha_r} \quad \text{for } r \in R$$

with $\sigma \geqq \varrho$ exactly as in the proof of Proposition 9.1.3 we get the

**Theorem.** *The diametral dimension of the power series space $\Lambda$ with $\varrho_0 = +\infty$, resp. $\varrho_0 < +\infty$, consists of all positive sequences of numbers $[\delta_r, R]$ for which the relation*

$$\sup\{\delta_r^{-1}\, q^{\alpha_r} : r \in R\} < +\infty$$

*holds for one. resp, each, number $q$ between 0 and 1.*

**9.3.2.** From the previous theorem it follows that all power series spaces

$$\Lambda_{\varrho_0} = \{[\xi_r, R] : \sum_R |\xi_r|\, \varrho^{\alpha_r} < +\infty \quad \text{for } 0 < \varrho < \varrho_0\}$$

corresponding to finite positive numbers $\varrho_0$ have the same diametral dimension since we can completely describe $\Delta_{\mathfrak{U}}(\Lambda_{\varrho_0})$ in terms of the sequence $[\alpha_r, R]$. This fact is not however surprising, because of the

**Proposition.** *For each sequence $[\alpha_r, R]$, all power series spaces $\Lambda_{\varrho_0}$ with $0 < \varrho_0 < +\infty$ are isomorphic to one another.*

*Proof.* The power series space $\Lambda_{\varrho_0}$ is mapped isomorphically onto $\Lambda_1$ by the diagonal transformation $D$ with

$$D[\xi_r, R] = [\xi_r\, \varrho_0^{\alpha_r}, R]\,.$$

**9.3.3.** We now show that the diametral dimension is a complete invariant of the power series space.

**Theorem 1.** *Each power series space $\Lambda$ with $\varrho_0 = +\infty$ is uniquely determined by its diametral dimension.*

*Proof.* Let $P$ be the collection of all sequences $[\delta_r^{-1}, R]$ with $[\delta_r, R] \in \Delta_{\mathfrak{U}}(\Lambda)$. We then have

$$[\varrho^{\alpha_r}, R] \in P \qquad \text{for all} \quad \varrho > 0\,.$$

On the other hand, for each sequence $[\varrho_r, R] \in P$, there is a number $q$ between 0 and 1 such that the relation

$$\sigma = \sup\{\varrho_r\, q^{\alpha_r} : r \in R\} < +\infty$$

is valid since $[\varrho_r^{-1}, R] \in \Delta_{\mathfrak{U}}(\Lambda)$. Hence, for $\varrho = q^{-1}$ we get the estimate

$$\varrho_r \leqq \sigma\, \varrho^{\alpha_r} \qquad \text{for } r \in R\,. \tag{*}$$

Thus we have shown that the set $P$ consists precisely of all positive numbers $[\varrho_r, R]$ for which (*) is satisfied for certain positive numbers $\varrho$ and $\sigma$. Therefore $\Lambda$ coincides with the sequence space $\Lambda(P)$.

**Theorem 2.** *Each power series space with $\varrho_0 = 1$ is uniquely determined by its diametral dimension.*

*Proof.* Let $P$ be the collection of all positive sequences of numbers $[\varrho_r, R]$ for which

$$\sup\{\delta_r^{-1}\, \varrho_r : r \in R\} < +\infty$$

always holds for $[\delta_r, R] \in \Delta_{\mathfrak{U}}(\Lambda)$. By Theorem 9.3.1 we then have

$$[\varrho^{\alpha_r}, R] \in P \quad \text{for} \quad 0 < \varrho < 1 .$$

We now show that $P$ consists precisely of all positive sequences of numbers $[\varrho_r, R]$ for which the relation

$$\varrho_r \leqq \sigma \varrho^{\alpha_r} \quad \text{for } r \in R \tag{1}$$

is satisfied for a positive number $\sigma$ and a number $\varrho$ between 0 and 1. For this purpose we set

$$R_n = \left\{ r \in R : \varrho_r > n \left(1 - \frac{1}{n}\right)^{\alpha_r} \right\} \quad \text{for} \quad n \in N .$$

We then have

$$R = R_1 \supset R_2 \supset \cdots \quad \text{and} \quad \bigcap_N R_n = \varnothing . \tag{2}$$

If we assume that all sets $R_r$ are infinite, there exists a sequence of distinct numbers $r_n \in R_n$. Then, however, the sequence $[\delta_r, R]$ with

$$\delta_{r_n} = \frac{1}{n} \varrho_{r_n} \quad \text{and} \quad \delta_r = 1 \quad \text{for} \quad r \neq r_1, r_2, \ldots$$

belongs to $\Delta_{\mathfrak{U}}(\Lambda)$. We can thus determine for each number $q$ between 0 and 1 a natural number $n_0$ with

$$q \leqq \left(1 - \frac{1}{n}\right) \quad \text{for} \quad n \geqq n_0 .$$

But then we get the estimates

$$\delta_{r_n}^{-1} q^{\alpha_{r_n}} \leqq n \varrho_{r_n}^{-1} \left(1 - \frac{1}{n}\right)^{\alpha_{r_n}} \leqq 1 \quad \text{for} \quad n \geqq n_0$$

and

$$\delta_r^{-1} q^{\alpha_r} \leqq 1 \quad \text{for} \quad r \neq r_1, r_2, \ldots .$$

Consequently,

$$\sup \{\delta_r^{-1} q^{\alpha_r} : r \in R\} < +\infty \quad \text{for} \quad 0 < q < 1 ,$$

and the sequence $[\delta_r, R]$ lies in $\Delta_{\mathfrak{U}}(\Lambda)$. Since $[\varrho_r, R] \in P$ we arrive at the false statement

$$\sup \{n : n \in N\} = \sup \{\delta_{r_n}^{-1} \varrho_{r_n} : n \in N\} < +\infty .$$

Because of this contradiction it follows that at least one set $R_n$ must be finite. On the basis of statement (2) this happens only if the sets $R_n$ are almost all empty. From the relation $R_n = \varnothing$ we get the desired estimate

$$\varrho_r \leqq n \left(1 - \frac{1}{n}\right)^{\alpha_r} \quad \text{for all} \quad r \in R .$$

This characterization of the sequences belonging to $P$ leads to the assertion that $\Lambda$ coincides with the sequence space $\Lambda(P)$.

From 9.3.2 and the previous Theorem we now get

**Theorem 3.** *Each power series space $\Lambda$ with $\varrho_0 < +\infty$ is uniquely determined up to a diagonal transformation by its diametral dimension.*

**9.3.4.** It is easy to see that all power series spaces for which

$$\lim_r \alpha_r < +\infty$$

is valid coincide with the Banach space $l_R^1$. Here the positive number $\varrho_0$ can be finite or infinite. If we neglect this trivial case we get the

**Theorem.** *Two power series spaces of different type can never be isomorphic.*

*Proof.* We consider two power series spaces $\Lambda_\alpha$ and $\Lambda_\beta$ corresponding to the systems

$$P_\alpha = \{[\varrho^{\alpha_r}, R] : 0 < \varrho < +\infty\}$$

and

$$P_\beta = \{[\varrho^{\beta_r}, R] : 0 < \varrho < 1\}.$$

Further, we suppose that

$$\lim_r \alpha_r = +\infty.$$

Now assume that the two spaces $\Lambda_\alpha$ and $\Lambda_\beta$ are isomorphic, so that the identity

$$\Delta_{\mathfrak{U}}(\Lambda_\alpha) = \Delta_{\mathfrak{U}}(\Lambda_\beta)$$

holds. Then the sequences

$$[n^{-\alpha_r}, R] \in \Delta_{\mathfrak{U}}(\Lambda_\alpha) \quad \text{with} \quad n \in N$$

also lie in $\Delta_{\mathfrak{U}}(\Lambda_\beta)$ and we have

$$\varrho_n = 1 + \sup\left\{n^{\alpha_r}\left(1 - \frac{1}{n}\right)^{\beta_r} : r \in R\right\} < +\infty.$$

From the inequality

$$\varrho_n^{-1}\, n^{\alpha_r} \leqq \left(1 - \frac{1}{n}\right)^{-\beta_r}$$

it follows that for each $r \in R$ the set of numbers $\varrho_n^{-1}\, n^{\alpha_r}$ with $n \in N$ is bounded. Consequently, we can define the numbers

$$\delta_r^{-1} = \sum_N 2^{-n}\, \varrho_n^{-1}\, n^{\alpha_r}.$$

For each number $q$ between 0 and 1 we determine a natural number $n_0$ with

$$q \leqq \left(1 - \frac{1}{n}\right) \quad \text{for} \quad n \geqq n_0.$$

Moreover, we note that since

$$[n^{-\alpha_r}, R] \in \Delta_{\mathfrak{U}}(\Lambda_\alpha) = \Delta_{\mathfrak{U}}(\Lambda_\beta),$$

the relations

$$\sigma_n = \sup \{n^{\alpha_r} q^{\beta_r} : r \in R\} < +\infty$$

hold. Therefore, for all $r \in R$ we have the estimate

$$\delta_r^{-1} q^{\beta_r} \leqq \sum_{n=n_0}^{\infty} 2^{-n} \varrho_n^{-1} n^{\alpha_r} \left(1 - \frac{1}{n}\right)^{\beta_r} + \sum_{n=1}^{n_0-1} 2^{-n} \varrho_n^{-1} n^{\alpha_r} q^{\beta_r},$$

from which we obtain the statement.

$$\sup \{\delta_r^{-1} q^{\beta_r} : r \in R\} \leqq 1 + \sum_{n=1}^{n_0-1} 2^{-n} \varrho_n^{-1} \sigma_n < +\infty .$$

It follows from this that the sequence $[\delta_r, R]$ belongs to $\Delta_{\mathfrak{U}}(\Lambda_\beta)$ and hence to $\Delta_{\mathfrak{U}}(\Lambda_\alpha)$ as well. Thus there exists a natural number $m$ with

$$\sigma = \sup \{\delta_r^{-1} m^{-\alpha_r} : r \in R\} < +\infty .$$

But this is impossible since

$$\delta_r^{-1} \geqq 2^{-2m} \varrho_{2m}^{-1} (2m)^{\alpha_r}$$

would then lead to the contradiction

$$2^{\alpha_r} \leqq 2^{2m} \varrho_{2m} \delta_r^{-1} m^{-\alpha_r} \leqq 2^{2m} \varrho_{2m} \sigma \quad \text{for } r \in R .$$

From the resulting contradiction we conclude that the two power series spaces $\Lambda_\alpha$ and $\Lambda_\beta$ cannot be isomorphic.

## 9.4. The Diametral Dimension of Nuclear Locally Convex Spaces

**9.4.1.** We now give a characterization of nuclear locally convex spaces with the help of diametral dimension.

**Theorem.** *A locally convex space $E$ is nuclear if and only if for some, resp. each, positive number $\lambda$ the assertion*

$$[(r+1)^{-\lambda}, R] \in \Delta_{\mathfrak{U}}(E)$$

*is valid.*

*Necessity.* If $U \in \mathfrak{U}(E)$ is a zero neighborhood of the nuclear locally convex space $E$ we can determine by Theorem 8.6.1 a zero neighborhood $V \in \mathfrak{U}(E)$ with $V < U$ such that the canonical mapping from $E(V)$ onto $E(U)$ is of type $\mathfrak{l}^p$. Here we can always assume that

$$\sum_R \alpha_r(E(V, U))^p \leqq 1$$

holds. Since

$$\delta_r(V, U) \leqq \alpha_r(E(V, U))$$

by Lemma 9.1.6 we get

$$(s+1)\,\delta_s(V,U)^p \leqq \sum_{r=0}^{s} \delta_r(V,U)^p \leqq 1 \quad \text{for } s = 0, 1, \ldots .$$

Consequently, we have

$$\delta_s(V,U) \leqq (s+1)^{-1/p} \quad \text{for } s = 0, 1, \ldots .$$

*Sufficiency.* We now consider a locally convex space $E$ for which

$$[(r+1)^{-\lambda}, R] \in \Delta_{\mathfrak{U}}(E)$$

for some positive number $\lambda$. We can then determine for each zero neighborhood $U = U_0 \in \mathfrak{U}(E)$ zero neighborhoods $U_1, \ldots, U_n \in \mathfrak{U}(E)$ with $U_n < \cdots < U_1 < U_0$ and

$$\delta_r(U_k, U_{k-1}) \leqq (r+1)^{-\lambda} \quad \text{for } r \in R .$$

Here the natural number $n$ is assumed so large that $n\lambda > 2$. If $\varepsilon$ is an arbitrary positive number, there exist for each number $r$ linear subspaces $F_k$ of $E$, whose dimension is at most $r$ and for which

$$U_k \subset ((r+1)^{-\lambda} + \varepsilon)\, U_{k-1} + F_k .$$

Consequently,

$$V = U_n \subset ((r+1)^{-\lambda} + \varepsilon)^n\, U + F .$$

Here $F = F_1 + \cdots + F_n$ is a linear subspace of $E$ of dimension at most $n\,r$. Therefore, the estimate

$$\delta_{nr}(V,U) \leqq ((r+1)^{-\lambda} + \varepsilon)^n$$

is valid, and by taking the limit as $\varepsilon \to 0$ we get

$$\delta_{nr}(V,U) \leqq (r+1)^{-n\lambda} \quad \text{for } r \in R .$$

Finally by applying Lemma 9.1.6 with $S = \{0, 1, 2, \ldots\}$ we get

$$\sum_S \alpha_s(E(V,U)) \leqq \sum_S (s+1)\,\delta_s(V,U) \leqq \sum_R \sum_{i=0}^{n-1} (n\,r + i + 1)\,\delta_{nr+i}(V,U)$$

$$\leqq n^2 \sum_R (r+1)\,\delta_{nr}(V,U) \leqq n^2 \sum_R (r+1)^{-n\lambda+1} < +\infty .$$

Consequently, the canonical mapping $E(V,U)$ is of type $\boldsymbol{l}^1$, and the locally convex space $E$ must be nuclear on the basis of Theorem 8.6.1.

**9.4.2.** By applying the same reasoning as in 9.4.1 we get the

**Theorem.** *A locally convex space $E$ is dual nuclear if and only if for some, resp. each, positive number $\lambda$ the assertion*

$$[(r+1)^{-\lambda}, R] \in \Delta_{\mathfrak{B}}(E)$$

*is valid.*

## 9.5. A Characterization of Dual Nuclear Locally Convex Spaces

**9.5.1.** For each null sequence $[x_r, R]$ of elements of a locally convex space $E$ we set

$$A[x_r, R] = \{x \in E : x = \sum_R \xi_r x_r \quad \text{with} \quad \sum_R |\xi_r| \leqq 1\} .$$

We can show that $A[x_r, R]$ is the smallest closed and absolutely convex subset of $E$ which contains all elements $x_r$.

**9.5.2. Theorem.** *A locally convex space $E$ is dual nuclear if and only if for each bounded subset $A$ there is a totally summable family $[x_r, R]$ from $E$ with*

$$A \subset A[x_r, R] .$$

*Necessity.* Without loss of generality we can assume that the set $A$ belongs to $\mathfrak{B}(E)$. Since the locally convex space $E$ is dual nuclear there is a set $B \in \mathfrak{B}(E)$ with $A < B$ such that the canonical mapping from $E(A)$ into $E(B)$ is of type $\mathfrak{l}^{1/2}$. On the basis of Proposition 8.4.2 there exist linear forms $\mathfrak{a}_r$ in the closed unit ball of $E(A)'$, elements $y_r \in B$ and numbers $\lambda_r$ with

$$1 + \sum_R |\lambda_r| = \lambda < +\infty ,$$

such that the relation

$$x = \sum_R \lambda_r^2 \langle x, \mathfrak{a}_r\rangle y_r \qquad \text{for all} \quad x \in E(A)$$

holds.

If we set $x_r = \lambda \lambda_r y_r$ it follows from the inequality

$$\sum_R p_B(x_n) \leqq \lambda^2 ,$$

that the family $[x_r, R]$ is totally summable. Moreover, each element $x \in A$ can be represented in the form

$$x = \sum_R \xi_r x_r \quad \text{with} \quad \xi_r = \lambda^{-1} \lambda_r \langle x, \mathfrak{a}_r\rangle \quad \text{and} \quad \sum_R |\xi_r| \leqq 1 .$$

*Sufficiency.* We consider an arbitrary set $A \in \mathfrak{B}(E)$ and for it determine a totally summable family $[x_r, R]$ with

$$A \subset A[x_r, R] .$$

Then for some set $B \in \mathfrak{B}(E)$

$$\sum_R p_B(x_r) \leqq 1$$

holds and by reordering $R$ we can assume that the sequence of numbers $p_B(x_r)$ is monotonically decreasing.

We now represent each element $x \in A$ as

$$x = \sum_R \xi_r x_r \quad \text{with} \quad \sum_R |\xi_r| \leqq 1$$

and set

$$y = \sum_{r=s}^{\infty} \xi_r x_r .$$

We then have

$$p_B(y) \leqq \sum_{r=s}^{\infty} |\xi_r| \, p_B(x_r) \leqq p_B(x_s) .$$

If the linear subspace of $E$ determined by the elements $x_0, \ldots, x_{s-1}$ is denoted by $F$, then from

$$x = y + \sum_{r=0}^{s-1} \xi_r x_r \in p_B(x_s) \, B + F$$

there follows

$$A \subset p_B(x_s) \, B + F .$$

Since $\dim F \leqq s$ the relation

$$\delta_s(A, B) \leqq p_B(x_s) \quad \text{for } s \in R$$

is thus proved. Therefore

$$(r + 1) \, \delta_r(A, B) \leqq \sum_{s=0}^{r} \delta_s(A, B) \leqq 1 \quad \text{for } r \in R .$$

Thus the sequence $[(r + 1)^{-1}, R]$ belongs to $\Delta_{\mathfrak{B}}(E)$ and the locally convex space $E$ is dual nuclear on the basis of Theorem 9.4.2.

**9.5.3.** A sequence $[x_r, R]$ of elements of a locally convex space $E$ is called **rapidly decreasing** if all sequences $[(r + 1)^n x_r, R]$ with $n = 1, 2, \ldots$ are bounded.

**Theorem.** *A quasi-complete locally convex space E is dual nuclear if and only if for each bounded subset A there is a rapidly decreasing sequence* $[x_r, R]$ *in E with*

$$A \subset A[x_r, R] .$$

*Necessity.* By 0.11.4 and Proposition 4.4.1 there exists in $E$ a fundamental system $\mathfrak{B}_{\mathfrak{H}}(E)$ of bounded subsets such that the normed spaces $E(A)$ with $A \in \mathfrak{B}_{\mathfrak{H}}(E)$ are Hilbert spaces. Here the completeness of $E(A)$ results from the quasi-completeness of $E$.

Without loss of generality we can assume that the bounded subset $A$ belongs to $\mathfrak{B}_{\mathfrak{H}}(E)$. Then for each natural number $n$ there is a set $B_n \in \mathfrak{B}_{\mathfrak{H}}(E)$ such that the canonical mapping from $E(A)$ into $E(B_n)$ is of type $l^{1/n}$. Hence, because of Theorem 8.3.2, there exist two orthonormal systems $[e_r^{[n]}, R]$ and $[f_r^{[n]}, R]$ of $E(A)$ resp. $E(B_n)$ with

$$x = \sum_R \alpha_r^{[n]} (x, e_r^{[n]})_A f_r^{[n]} \quad \text{for } x \in E(A),$$

$$\alpha_0^{[n]} \geqq \alpha_1^{[n]} \geqq \cdots \geqq 0 \quad \text{and} \quad \left\{ \sum_R |\alpha_r^{[n]}|^{1/n} \right\}^n = \varrho_n < +\infty .$$

Consequently,

$$(r+1)^n \alpha_r^{[n]} \leqq \varrho_n \quad \text{for } r \in R .$$

Since

$$(x, e_r^{[n]})_A = 0 \quad \text{for } r \in R$$

always implies $x = o$, the orthonormal systems $[e_r^{[n]}, R]$ are complete for all natural numbers $n$.

By orthonormalizing the sequence

$$e_1^{[1]}, e_1^{[2]}, e_2^{[2]}, e_2^{[1]}, e_1^{[3]}, e_2^{[3]}, \ldots$$

(usual counting procedure for the positive rational numbers) we get in $E(A)$ a complete orthonormal system $[e_r, R]$ with

$$(e_s, e_r^{[n]})_A = 0 \quad \text{for } s > \max(n^2, r^2)$$

Therefore the identity

$$e_s = \sum_{r^2 \geqq s} (e_s, e_r^{[n]})_A e_r^{[n]} \quad \text{for } s > n^2$$

is valid, and from it as well as

$$e_r^{[n]} = \alpha_r^{[n]} f_r^{[n]}$$

we get the estimate

$$p_{B_n}(e_s)^2 = \sum_{r^2 \geqq s} |(e_s, e_r^{[n]})_A|^2 |\alpha_r^{[n]}|^2 \leqq (s+1)^{-n} \varrho_n^2 \quad \text{for } s > n^2 .$$

Consequently,

$$(s+1)^{n/2} e_s \in \varrho_n B_n \quad \text{for } s > n^2 .$$

We have thus shown that the sequence

$$x_r = \varrho (r+1) e_r \quad \text{with } \varrho = \left\{ \sum_R (r+1)^2 \right\}^{1/2}$$

is rapidly decreasing. Finally, from the identity

$$x = \sum_R \xi_r x_s \quad \text{with } \xi_r = \varrho^{-1} (r+1)^{-1} (x, e_r)_A$$

we obtain the desired relation

$$A \subset A[x_r, R] .$$

The *sufficiency* of the given condition follows immediately from Theorem 9.5.2 because each rapidly decreasing sequence is totally summable.

## 9.6. The $\varepsilon$-Content of Bounded Subsets in Normed Spaces

**9.6.1.** Let $B$ be an arbitrary bounded subset of a normed space $E$ with closed unit ball $U$. For each positive number $\varepsilon$ the supremum $M_\varepsilon(B)$ of all natural numbers $m$, for which there exist elements

$$x_1, \ldots, x_m \in B \quad \text{with } x_i - x_k \notin \varepsilon U \quad \text{for } i \neq k$$

is designated as the **$\varepsilon$-content** of $B$. Thus $M_\varepsilon(B)$ is either a natural number or $+\infty$ and we have

$$M_{\varepsilon'}(B) \geqq M_\varepsilon(B) \quad \text{for } 0 < \varepsilon' < \varepsilon .$$

**9.6.2.** The concept of $\varepsilon$-content is really only meaningful for precompact subsets as is shown by the following

**Proposition.** *A bounded subset $B$ of a normed space $E$ is precompact if and only if*

$$M_\varepsilon(B) < +\infty$$

*is valid for all positive numbers $\varepsilon$.*

*Proof.* If the set $B$ is precompact there are for each positive number $\varepsilon$ finitely many elements $y_1, \ldots, y_n \in E$ with

$$B \subset \bigcup_{s=1}^{n} \left\{y_s + \frac{1}{2} \varepsilon U\right\}.$$

We then have

$$M_\varepsilon(B) \leqq n .$$

In fact, if there were $n+1$ elements $x_1, \ldots, x_{n+1} \in B$ with $x_i - x_k \notin \varepsilon U$ for $i \neq k$ there exist at least two different elements $x_i$ and $x_k$ contained in the same set $y_s + \frac{1}{2} \varepsilon U$. But this is impossible because it would lead to the false statement $x_i - x_k \in \varepsilon U$.

We now consider a bounded set $B$ with

$$M_\varepsilon(B) < +\infty \quad \text{for } \varepsilon > 0$$

and determine for the natural number $m = M_\varepsilon(B)$, elements $x_1, \ldots, x_m \in B$ with $x_i - x_k \notin \varepsilon U$ for $i \neq k$. If $x$ is an arbitrary element from $B$ then for at least one $x_i$ the assertion $x \in x_i + \varepsilon U$ must be valid since otherwise, the system $\{x_1, \ldots, x_m\}$ could be enlarged by addition of the element $x$ and that would lead to the false relation $M_\varepsilon(B) \geqq m + 1$. Consequently,

$$B \subset \bigcup_{i=1}^{m} \{x_i + \varepsilon U\} .$$

**9.6.3.** In the following two lemmas we determine the relation between the diameters of a bounded subset $B$ of a real normed space $E$ which was defined in 9.1.1 and the $\varepsilon$-content.

**Lemma 1.** *For each bounded subset $B$ of a real normed space $E$*

$$\delta_r(B) < \frac{1}{3}\varepsilon$$

*implies the inequality*

$$M_\varepsilon(B) \leqq (3\,\delta_0(B)\,\varepsilon^{-1} + 2)^r .$$

*Proof.* By hypothesis there is a linear subspace $F$ of $E$ with

$$B \subset \frac{1}{3}\varepsilon\, U + F \quad \text{and} \quad s = \dim F \leqq r .$$

We now consider finitely many elements $x_1, \ldots, x_m \in B$ with $x_i - x_k \notin \varepsilon\, U$ for $i \neq k$ and represent them in the form

$$x_i = \frac{1}{3}\varepsilon\, y_i + z_i \quad \text{with } y_i \in U \text{ and } z_i \in F .$$

Then, since

$$z_i = x_i - \frac{1}{3}\varepsilon\, y_i \in B + \frac{1}{3}\varepsilon\, U$$

all elements $z_1, \ldots, z_m$ lie in the set

$$B_0 = \left\{B + \frac{1}{3}\varepsilon\, U\right\} \cap F .$$

Since the assumption

$$z_i - z_k \in \frac{2}{3}\varepsilon\, U \quad \text{for } i \neq k$$

leads to the contradiction

$$x_i - x_k = \frac{1}{3}\varepsilon(y_i - y_k) + (z_i - z_k) \in \varepsilon\, U ,$$

we must have

$$z_i - z_k \notin \frac{2}{3}\varepsilon\, U \quad \text{for } i \neq k .$$

Consequently, the closed sets

$$S_i = z_i + \frac{1}{3}\varepsilon\, S \quad \text{with } S = U \cap F$$

are disjoint. Moreover, since $B \subset \delta_0(B)\, U$, the relation

$$S_i \subset \left(B_0 + \frac{1}{3}\varepsilon\, S\right) \subset \left(B + \frac{2}{3}\varepsilon\, U\right) \cap F \subset \left(\delta_0(B) + \frac{2}{3}\varepsilon\right) S$$

holds.

We now introduce a measure $\mu$ on the $s$-dimensional real linear space $F$ by mapping the latter one-to-one and linearly onto the $s$-dimensional

Euclidean space and transferring Lebesgue measure $\mu$ defined there. We then have the estimate

$$m\left(\frac{1}{3}\varepsilon\right)^s \mu(S) = \sum_{i=1}^{m} \mu(S_i) \leqq \left(\delta_0(B) + \frac{2}{3}\varepsilon\right)^s \mu(S),$$

from which we get

$$m \leqq (3\,\delta_0(B)\,\varepsilon^{-1} + 2)^s$$

since $\mu(S) > 0$. The desired inequality

$$M_\varepsilon(B) \leqq (3\,\delta_0(B)\,\varepsilon^{-1} + 2)^r$$

is thus proved because $s \leqq r$.

**Lemma 2.** *For each absolutely convex bounded subset $B$ of a real normed space $E$ the inequality*

$$\delta_0(B) \cdots \delta_r(B) \leqq (r+1)!\,\varepsilon^{r+1}\,M_\varepsilon(B)$$

*is valid for all numbers $r = 0, 1, \ldots$ and $\varepsilon > 0$.*

*Proof.* Since the assertion is trivial for $\delta_r(B) = 0$, we need only treat the case $\delta_r(B) > 0$. For this purpose we consider arbitrary numbers $\delta_0, \ldots, \delta_r$ with $0 < \delta_s < \delta_s(B)$ and determine recursively elements $x_0, \ldots, x_r \in B$ such that

$$x_s \notin \delta_s U + E_s \quad \text{for } s = 0, \ldots, r.$$

Here $E_s$ is assumed to be the $s$-dimensional subspace of $E$ determined by the elements $x_i$ with $i < s$.

For $\varepsilon$ an arbitrary positive number we consider the collection of all elements

$$y(\alpha_0, \ldots, \alpha_r) = \varepsilon \sum_{s=0}^{r} \alpha_s\,\delta_s^{-1}\,x_s,$$

where the coefficients $\alpha_0, \ldots, \alpha_r$ are integers. We then get

$$y(\alpha_0, \ldots, \alpha_r) - y(\beta_0, \ldots, \beta_r) \notin \varepsilon U \quad \text{for } (\alpha_0, \ldots, \alpha_r) \neq (\beta_0, \ldots, \beta_r).$$

In fact, let us assume the above relation does not hold and set

$$k = \sup\{s : \alpha_s \neq \beta_s\}.$$

Then the statement

$$\varepsilon(\alpha_k - \beta_k)\,\delta_k^{-1}\,x_k = y(\alpha_0, \ldots, \alpha_r) - y(\beta_0, \ldots, \beta_r) - \varepsilon \sum_{s=0}^{k-1} (\alpha_s - \beta_s)\,\delta_s^{-1}\,x_s$$

leads to

$$\varepsilon(\alpha_k - \beta_k)\,\delta_k^{-1}\,x_k \in \varepsilon U + E_k$$

and we get a contradiction

$$x_k \in (\alpha_k - \beta_k)^{-1}\,\delta_k U + E_k \subset \delta_k U + E_k.$$

We now determine a lower bound for the number $m$ of elements $y(\alpha_0, \ldots, \alpha_r)$ contained in the set

$$S = \left\{ \sum_{s=0}^{r} \xi_s x_s : \sum_{s=0}^{r} |\xi_s| \leqq 1 \right\}.$$

For this purpose we first note that the sets

$$Q(\alpha_0, \ldots, \alpha_r) = \left\{ \varepsilon \sum_{s=0}^{r} (\alpha_s + \eta_s) \, \delta_s^{-1} x_s : |\eta_s| \leqq 1 \right\}$$

form a covering of $S$. If the elements $x_0, \ldots, x_r$ are considered as unit vectors, we can identify $E_{r+1}$ with $(r+1)$-dimensional Euclidean space. With respect to Lebesgue measure $\mu$ transferred to $E_{r+1}$ the following statements are valid:

$$\mu(S) = \frac{2^{r+1}}{(r+1)!} \quad \text{and} \quad \mu(Q(\alpha_0, \ldots, \alpha_r)) = (2\,\varepsilon)^{r+1} (\delta_0 \ldots \delta_r)^{-1}.$$

Since

$$\mu(S) \leqq \sum \mu(Q(\alpha_0, \ldots, \alpha_r))$$

we get

$$\frac{2^{r+1}}{(r+1)!} \leqq m \, (2\,\varepsilon)^{r+1} (\delta_0 \ldots \delta_r)^{-1}.$$

Since the bounded set $B$ is absolutely convex by hypothesis $S \subset B$. Consequently, there are in $B$ at least $m$ distinct elements $y(\alpha_0 \ldots, \alpha_r)$ and we have $m \leq M_\varepsilon(B)$. Therefore,

$$\delta_0 \cdot \cdot \cdot \delta_r \leqq (r+1)! \, \varepsilon^{r+1} M_\varepsilon(B),$$

and we obtain the desired inequality by taking the limit as $\delta_s \to \delta_s(B)$.

**9.6.4.** As the first consequence of the inequalities proved in 9.6.3 we get the

**Lemma.** *For an absolutely convex and bounded subset $B$ of a real normed space $E$ the assertions*

$$\delta_s(B) = 0 \quad \textit{and} \quad \lim_{\varepsilon \to 0} \varepsilon^\lambda M_\varepsilon(B) = 0 \qquad \textit{for } \lambda > s$$

*are equivalent.*

*Proof.* If $\delta_s(B) = 0$ then the inequality

$$M_\varepsilon(B) \leqq (3\,\delta_0(B)\,\varepsilon^{-1} + 2)^s$$

holds for all positive numbers $\varepsilon$ by Lemma 1, and from it we get

$$\lim_{\varepsilon \to 0} \varepsilon^\lambda M_\varepsilon(B) = 0 \quad \text{for } \lambda > s.$$

On the other hand, if the limit condition is satisfied we obtain from the estimate
$$\delta_0(B) \cdots \delta_s(B) \leqq (s+1)!\, \varepsilon^{s+1} M_\varepsilon(B)$$
of Lemma 2 the assertion
$$\delta_0(B) \cdots \delta_s(B) = 0$$
by taking the limit as $\varepsilon \to 0$. Therefore, $\delta_s(B) = o$.

## 9.7. The Approximative Dimension of Locally Convex Spaces

**9.7.1.** In an arbitrary locally convex space $E$ we consider two zero neighborhoods $U, V \in \mathfrak{U}(E)$ with $V < U$. For each positive number $\varepsilon$ the supremum $M_\varepsilon(V, U)$ of all natural numbers $m$ for which there are elements $x_1, \ldots, x_m \in V$ with $x_i - x_k \notin \varepsilon U$ for $i \neq k$ is designated as the $\varepsilon$-**content of** $V$ **with respect to** $U$. It can be shown that $M_\varepsilon(V, U)$ coincides with the $\varepsilon$-content of the bounded subset
$$V(U) = \{x(U) : x \in V\}$$
of the normed space $E(U)$.

**9.7.2.** The **approximative dimension** $\Phi_{\mathfrak{U}}(E)$ of a locally convex space $E$ is the collection of all positive functions $\varphi = [\varphi(\varepsilon)]$ defined on the interval $(0, +\infty)$ which have the property that for each zero neighborhood $U \in \mathfrak{U}(E)$ there is a zero neighborhood $V \in \mathfrak{U}(E)$ with $V < U$ such that
$$\lim_{\varepsilon \to 0} \varphi(\varepsilon)^{-1} M_\varepsilon(V, U) = 0\,.$$

It is easy to see that we can replace $\mathfrak{U}(E)$ by an arbitrary fundamental system of zero neighborhoods in this definition.

**9.7.3.** By means of the same considerations as in 9.2.3 we get the

**Theorem.** *For two isomorphic locally convex spaces $E$ and $F$ we have*
$$\Phi_{\mathfrak{U}}(E) = \Phi_{\mathfrak{U}}(F)\,.$$

**9.7.4.** The approximative dimension of a locally convex space is a meaningful generalization of the ordinary dimension as is shown by the

**Theorem.** *A real locally convex space $E$ is at most $s$-dimensional if and only if for all numbers $\lambda > s$ the function $[\varepsilon^{-\lambda}]$ belongs to $\Phi_{\mathfrak{U}}(E)$.*

*Proof.* If $U, V \in \mathfrak{U}(E)$ are two zero neighborhoods with $V < U$, then
$$V(U) = \{x(U) : x \in V\}$$

is a bounded subset of the normed space $E(U)$ for which the identities

$$\delta_s(V, U) = \delta_s(V(U)) \quad \text{and} \quad M_\varepsilon(V, U) = M_\varepsilon(V(U))$$

hold. Therefore the statements

$$\delta_s(U, V) = 0 \quad \text{and} \quad \lim_{\varepsilon \to 0} \varepsilon^\lambda M_\varepsilon(V, U) = 0 \quad \text{for } \lambda > s$$

are equivalent by Lemma 9.6.4, and the assertion follows from Theorem 9.2.4.

*Remark.* In the case of a complex locally convex space we have to replace $s$ by $2s$.

**9.7.5.** The approximative dimension $\Phi_{\mathfrak{U}}(E)$ is indeed a measure for the size of the locally convex space $E$ as is seen from the two following Propositions.

**Proposition 1.** *For each linear subspace $F$ of a locally convex space $E$,*

$$\Phi_{\mathfrak{U}}(F) \supset \Phi_{\mathfrak{U}}(E) .$$

*Proof.* If $\varphi$ is an arbitrary function from $\Phi_{\mathfrak{U}}(E)$, there is for each zero neighborhood $U \in \mathfrak{U}(E)$ a zero neighborhood $V \in \mathfrak{U}(E)$ with $V < U$, such that

$$\lim_{\varepsilon \to 0} \varphi(\varepsilon)^{-1} M_\varepsilon(V, U) = 0 .$$

But since for $m$ arbitrary elements $x_1, \ldots, x_m \in V_0 = F \cap V$ with $x_i - x_k \notin \varepsilon\, U_0 = \varepsilon\, (F \cap U)$ for $i \neq k$ we also have $x_i - x_k \notin \varepsilon\, U$ for $i \neq k$, we must have

$$M_\varepsilon(V_0, U_0) \leqq M_\varepsilon(V, U) .$$

Therefore,

$$\lim_{\varepsilon \to 0} \varphi(\varepsilon)^{-1} M_\varepsilon(V_0, U_0) = 0$$

also holds. If we note that the sets $U_0 = F \cap U$ with $U \in \mathfrak{U}(E)$ form a fundamental system of zero neighborhoods in $F$ we get the assertion $\varphi \in \Phi_{\mathfrak{U}}(F)$, and the proof is complete.

**Proposition 2.** *For each closed linear subspace $F$ of a locally convex space $E$,*

$$\Phi_{\mathfrak{U}}(E/F) \supset \Phi_{\mathfrak{U}}(E) .$$

*Proof.* Since the sets

$$\overline{U(F)} = \{x(F) \in E/F : \alpha\, x \in U + F \quad \text{for } |\alpha| < 1\} \quad \text{with } U \in \mathfrak{U}(E)$$

form a fundamental system of zero neighborhoods in the quotient space $E/F$, the proposition follows from the inequality

$$M_\varepsilon(\overline{V(F)}, \overline{U(F)}) \leqq M_\varepsilon(V, U)$$

which holds for two zero-neighborhoods $U$, $V \in \mathfrak{U}(E)$ with $V < U$.

We consider $m$ equivalence classes $x_1(F), \ldots, x_m(F)$ from $\overline{V(F)}$ with $x_i(F) - x_k(F) \notin \varepsilon\, \overline{U(F)}$ for $i \neq k$. On the basis of the definition of $\overline{U(F)}$ there exist numbers $\alpha_{ik}$ between 0 and 1 for which

$$\alpha_{ik}(x_i - x_k) \notin \varepsilon\, U + F .$$

Consequently, for the maximum $\alpha < 1$ of the $\alpha_{ik}$ the relation

$$\alpha(x_i - x_k) \notin \varepsilon\, U + F$$

certainly holds. Moreover, we have

$$\alpha\, x_i \in V + F .$$

Hence, the elements $\alpha\, x_i$ can be represented in the form

$$\alpha\, x_i = y_i + z_i \quad \text{with} \quad y_i \in V \quad \text{and} \quad z_i \in F .$$

Here $y_i - y_k \notin \varepsilon\, U$ for $i \neq k$ because the contrary would lead to the false statement

$$\alpha(x_i - x_k) = (y_i - y_k) + (z_i - z_k) \in \varepsilon\, U + F .$$

Thus we have found $m$ elements $y_1, \ldots, y_m \in V$ with $y_i - y_k \notin \varepsilon\, U$ for $i \neq k$. Consequently, $M_\varepsilon(V, U)$ can never be smaller than $M_\varepsilon(\overline{V(F)}, \overline{U(F)})$.

**9.7.6. Problem.** *Under what hypotheses on the locally convex space E does the relation*

$$\Phi_{\mathfrak{U}}(E'_b) = \Phi_{\mathfrak{U}}(E)$$

*hold?*

**9.7.7.** The **dual approximative dimension** $\Phi_{\mathfrak{B}}(E)$ of a locally convex space $E$ is the collection of all positive functions $\varphi = [\varphi(\varepsilon)]$, defined on the interval $(0, +\infty)$ which have the property that for each set $A \in \mathfrak{B}(E)$ there is a set $B \in \mathfrak{B}(E)$ with $A < B$ such that

$$\lim_{\varepsilon \to 0} \varphi(\varepsilon)^{-1} M_\varepsilon(A, B) = 0 .$$

Here the $\varepsilon$-content $M_\varepsilon(A, B)$ of $A$ with respect to $B$ is defined exactly as in 9.7.1.

It is not yet known whether the dual approximative dimension always coincides with the approximative dimension of the strong topological dual.

**Problem.** *Under what hypotheses on the locally convex space E does the relation*

$$\Phi_{\mathfrak{B}}(E) = \Phi_{\mathfrak{U}}(E'_b)$$

*hold?*

## 9.8. The Approximative Dimension of Nuclear Locally Convex Spaces

**9.8.1.** In an arbitrary locally convex space $E$ we consider two zero neighborhoods $V$, $U \in \mathfrak{U}(E)$ with $V < U$ and designate as the **order of** $V$ **with respect to** $U$ the infimum $\varrho(V, U)$ of all positive numbers $\varrho$, for which there is a positive number $\varepsilon_0$ such that

$$M_\varepsilon(V, U) \leqq \exp(\varepsilon^{-\varrho}) \quad \text{for} \quad 0 < \varepsilon < \varepsilon_0 .$$

If no number $\varrho$ with the given property exists we set $\varrho(V, U) = +\infty$. We then easily see that

$$\varrho(V, U) = \limsup_{\varepsilon \to 0} \frac{\ln \ln M_\varepsilon(V, U)}{\ln \varepsilon^{-1}} .$$

**9.8.2. Lemma.** *For three zero neighborhoods $U$, $V$, $W \in \mathfrak{U}(E)$ with $W < V < U$ we have*

$$\frac{1}{\varrho(W, U)} \geqq \frac{1}{\varrho(W, V)} + \frac{1}{\varrho(V, U)} .$$

*Proof.* We restrict ourselves to the case in which the orders $\varrho(V, U)$ and $\varrho(W, V)$ are finite and consider arbitrary numbers $\varrho, \varrho', \sigma, \sigma'$ with

$$\varrho > \varrho' > \varrho(V, U) \quad \text{and} \quad \sigma > \sigma' > \varrho(W, V) .$$

If we set

$$\alpha = \sigma(\varrho' + \sigma)^{-1} \quad \text{and} \quad \beta = \varrho(\varrho' + \sigma)^{-1} ,$$

we then have $\alpha \varrho = \beta \sigma$ and $\alpha + \beta > 1$. We now determine a positive number $\varepsilon_0$ such that for all numbers $\varepsilon$ with $0 < \varepsilon < \varepsilon_0$ the following inequalities hold:

$$2\,\varepsilon^{\alpha+\beta} \leqq \varepsilon , \qquad 2\,\varepsilon^{-\varrho\sigma(\varrho'+\sigma)^{-1}} \leqq \varepsilon^{-\varrho\sigma(\varrho'+\sigma')^{-1}} ,$$

$$M_{\varepsilon\alpha}(V, U) \leqq \exp(\varepsilon^{-\alpha\varrho}) , \qquad M_{\varepsilon\beta}(W, V) \leqq \exp(\varepsilon^{-\beta\sigma}) .$$

For a fixed number $\varepsilon$ with $0 < \varepsilon < \varepsilon_0$ we consider a maximal system of element

$$x_1, \ldots, x_r \in V \quad \text{with} \quad x_i - x_k \notin \varepsilon^\alpha U \quad \text{for} \quad i \neq k$$

and

$$y_1, \ldots, y_s \in W \quad \text{with} \quad y_i - y_k \notin \varepsilon^\beta V \quad \text{for} \quad i \neq k .$$

We then have

$$V \subset \bigcup_{i=1}^{r} \{x_i + \varepsilon^\alpha U\} \quad \text{and} \quad W \subset \bigcup_{k=1}^{s} \{y_k + \varepsilon^\beta V\} .$$

Consequently, the relation

$$W \subset \bigcup_{i=1}^{r} \bigcup_{k=1}^{s} \{\varepsilon^\beta x_i + y_k + \varepsilon^{\alpha+\beta} U\}$$

holds.

Thus, at most $r \cdot s$ elements

$$z_1, \ldots, z_t \in W \quad \text{with} \quad z_p - z_q \notin \varepsilon U \quad \text{for} \quad p \neq q$$

can exist, for otherwise at least two different elements $z_p$ and $z_q$ would have to lie in the same set

$$\varepsilon^\beta x_i + y_k + \varepsilon^{\alpha+\beta} U$$

and this would lead to contradiction

$$z_p - z_q \in 2\, \varepsilon^{\alpha+\beta}\, U \subset \varepsilon\, U\,.$$

Therefore, the estimate

$$M_\varepsilon(W, U) \leqq r \cdot s = M_{\varepsilon\alpha}(V, U) \cdot M_{\varepsilon\beta}(W, V) \leqq \exp\,(\varepsilon^{-\alpha\varrho} + \varepsilon^{-\beta\sigma})$$
$$\leqq \exp\,(2\,\varepsilon^{-\varrho\sigma(\varrho'+\sigma)^{-1}}) \leqq \exp\,(\varepsilon^{-\varrho\sigma(\varrho'+\sigma')^{-1}})$$

resp.

$$\varrho(W, U) \leqq \varrho\, \sigma(\varrho' + \sigma')^{-1}$$

holds and we obtain the lemma by taking the limit as $\varrho \to \varrho(V, U)$ and $\sigma \to \varrho(W, V)$.

**9.8.3.** After these preliminaries we get the

**Theorem.** *A locally convex space E is nuclear if and only if for some, resp. each, positive number $\varrho$ the following condition is satisfied:*

*For each zero neighborhood $U \in \mathfrak{U}(E)$ there is a zero neighborhood $V \in \mathfrak{U}(E)$ with $V < U$ and $\varrho(V, U) \leqq \varrho$.*

*Proof.* Each complex locally convex space $E$ can also be considered in a natural way as a real locally convex space. But since the nuclearity of $E$ and the numbers $\varrho(V, U)$ can be defined without reference to the scalar field, we may restrict ourselves to the consideration of real locally convex spaces in the following.

First we shall show that the given condition is satisfied for all positive numbers $\varrho$ if it is valid for one positive number $\varrho_0$. For this purpose we determine for an arbitrary given zero neighborhood $U = U_0 \in \mathfrak{U}(E)$ zero neighborhoods $U_1, \ldots, U_n \in \mathfrak{U}(E)$ with

$$U_k < U_{k-1} \quad \text{and} \quad \varrho(U_k, U_{k-1}) \leqq \varrho_0 \quad \text{for} \quad k = 1, \ldots, n$$

and set $V = U_n$. If we choose the natural number $n$ so large that $n\,\varrho \geqq \varrho_0$ is true we get the inequality

$$\frac{1}{\varrho(V, U)} \geqq \sum_{k=1}^{n} \frac{1}{\varrho(U_k, U_{k-1})} \geqq \frac{n}{\varrho_0} \geqq \frac{1}{\varrho}\,.$$

Thus we have $\varrho(V, U) \leqq \varrho$, and the assertion is proved.

*Necessity.* By Theorem 9.4.1 there is for each zero neighborhood in a nuclear locally convex space $E$ a zero neighborhood $V \in \mathfrak{U}(E)$ with $U < V$ and

$$\delta_r(V, U) \leqq (r+1)^{-1} \quad \text{for } r \in R .$$

For an arbitrary positive number $\varepsilon$ we now determine that number $s \in R$ for which

$$3\,\varepsilon^{-1} - 1 < s \leqq 3\,\varepsilon^{-1} .$$

We then have

$$\delta_s(V, U) \leqq (s+1)^{-1} < \frac{1}{3}\,\varepsilon ,$$

and by Lemma 1 from 9.6.3 the inequality

$$M_\varepsilon(V, U) \leqq (3\,\varepsilon^{-1} + 2)^s \leqq (3\,\varepsilon^{-1} + 2)^{3\,\varepsilon^{-1}} \quad \text{for all } \varepsilon > 0$$

is valid. If a positive number $\varepsilon_0$ is now determined such that

$$(3\,\varepsilon^{-1} + 2)^3 \leqq \exp(\varepsilon^{-1}) \quad \text{for } 0 < \varepsilon < \varepsilon_0$$

we then have

$$M_\varepsilon(V, U) \leqq \exp(\varepsilon^{-2}) \quad \text{for } 0 < \varepsilon < \varepsilon_0 \quad \text{or} \quad \varrho(V, U) \leqq 2 .$$

We have thus shown that the given condition is satisfied for the integer 2.

*Sufficiency.* We now consider a locally convex space $E$ in which there is for each zero neighborhood $U \in \mathfrak{U}(E)$ a zero neighborhood $V \in \mathfrak{U}(E)$ with $V < U$ and $\varrho(V, U) \leqq 1/3$. Then there exists a positive number $\varepsilon_0$ with

$$M_\varepsilon(V, U) \leqq \exp(\varepsilon^{-1/3}) \quad \text{for } 0 < \varepsilon < \varepsilon_0 .$$

If $r_0$ is a natural number with $(r_0 + 1)^2\,\varepsilon_0 \geqq 1$ then

$$\delta_r(V, U) \leqq e\,(r+1)^{-1} \quad \text{for all } r \geqq r_0 .$$

In fact if we assume that there is a natural number $s \geqq r_0$ with

$$\delta_s(V, U) > e\,(s+1)^{-1}$$

we obtain the inequality

$$e^{s+1}\,(s+1)^{-(s+1)} \leqq \delta_0(V, U) \cdots \delta_s(V, U) \leqq (s+1)!\,\varepsilon^{s+1}\,M_\varepsilon(V, U)$$

on the basis of Lemma 2 of 9.6.3. If we set

$$\varepsilon = (s+1)^{-2} ,$$

then the estimates

$$M_\varepsilon(V, U) \leqq \exp\left((s+1)^{2/3}\right) \quad \text{and} \quad (s+1)! \leqq (s+1)^{s+1}$$

together with multiplication by $(s+1)^{s+1}$ and taking natural logarithmus lead to the contradiction

$$s + 1 \leqq (s+1)^{2/3} .$$

Finally we determine a positive number $\sigma \leqq e^{-1}$ such that

$$\sigma \, \delta_r(V, U) \leqq (r+1)^{-1} \quad \text{for} \quad r = 0, \ldots, r_0 - 1 \,.$$

We then have

$$\delta_r(\sigma \, V, U) \leqq (r+1)^{-1} \quad \text{for all} \quad r \in R \,,$$

and from Theorem 9.4.1 it follows that the space $E$ under consideration is nuclear.

**9.8.4.** As an immediate consequence of 9.8.3 we get the

**Theorem.** *A locally convex space $E$ is nuclear if and only if for some, resp. each, positive number $\varrho$*

$$[\exp(\varepsilon^{-\varrho})] \in \Phi_{\mathfrak{U}}(E) \,.$$

**9.8.5.** If we define the order $\varrho(A, B)$ of two sets $A, B \in \mathfrak{B}(E)$ just as in 9.8.1, we obtain the

**Theorem.** *A locally convex space $E$ is dual nuclear if and only if for some, resp. each, positive number $\varrho$ the following assertion is valid:*

*For each set $A \in \mathfrak{B}(E)$ there is a set $B \in \mathfrak{B}(E)$ with $A < B$ and $\varrho(A, B) \leqq \varrho$.*

**9.8.6.** As an analog to Theorem 9.8.4 we formulate the

**Theorem.** *A locally convex space $E$ is dual nuclear if and only if for some, resp. each, positive number $\varrho$*

$$[\exp(\varepsilon^{-\varrho})] \in \Phi_{\mathfrak{B}}(E) \,.$$

**9.8.7.** Let $E$ be an arbitrary locally convex space. For each set $K \in \mathfrak{K}(E)$ and all zero neighborhoods $U \in \mathfrak{U}(E)$ we can define the $\varepsilon$-content $M_\varepsilon(K, U)$ to be the supremum of all natural numbers $m$ for which there are elements

$$x_1, \ldots, x_m \in K \quad \text{with} \quad x_i - x_k \notin \varepsilon \, U \quad \text{for} \quad i \neq k \,.$$

The $\varepsilon$-content $M_\varepsilon(K, U)$ is always finite. If we put

$$\varrho(K, U) = \limsup_{\varepsilon \to 0} \frac{\ln \ln M_\varepsilon(K, U)}{\ln \varepsilon^{-1}}$$

as in 9.8.1, then by applying Theorem 9.8.3 we can easily show that $\varrho(K, U) = 0$. The same assertion holds for dual nuclear locally convex spaces.

Without proof we mention the

**Theorem.** *A metric or dual metric locally convex space $E$ is nuclear if and only if $\varrho(K, U) = 0$ for $K \in \mathfrak{K}(E)$ and $U \in \mathfrak{U}(E)$.*

Chapter 10

# Nuclear Locally Convex Spaces with Basis

A sequence of elements $e_0, e_1, \ldots$ of a locally convex space $E$ is called a *basis* if for each element $x \in E$ there is a unique sequence of numbers $\xi_0, \xi_1, \ldots$ for which

$$x = \lim_s \sum_{r=0}^{s} \xi_r e_r$$

holds. In general, the order of the elements $e_0, e_1, \ldots$ can not be varied. We cite as an example the Banach space $\mathscr{C}[0, 1]$ for which a basis was constructed by J. Schauder but in which no basis exists for which all series

$$\sum_R \xi_r e_r$$

converge unconditionally. The case in which the series

$$\sum_R \xi_r e_r$$

converge absolutely is encountered even less frequently among Banach spaces. In fact, it follows, from Theorem 10.1.4 that these spaces are isomorphic to the sequence space $\boldsymbol{l}_R^1$.

In view of this fact it is surprising that for each basis in a nuclear $(F)$-space all series

$$\sum_R \xi_r e_r$$

converge absolutely, as shown by A. S. Dynin and B. S. Mitiagin. From this fundamental Basis Theorem, which we shall prove below in a very general form, it follows in conjunction with a method due to S. Rolewicz [1] that all nuclear complete locally convex spaces with an equicontinuous basis are isomorphic to nuclear sequence spaces (Theorem 10.2.2). Therefore, the question of general isomorphism criteria for nuclear sequence spaces appears as the most important problem with regard to the desired classification of nuclear locally convex spaces (10.1.3).

In Section 10.3 we exhibit bases in some concrete locally convex spaces. However, just as for separable Banach spaces, it is still unknown whether a basis exists in every nuclear $(F)$-space.

## 10.1. Locally Convex Spaces with Basis

**10.1.1.** A sequence $[e_r, R]$ of elements in a locally convex space $E$ is called **basis** if for each element $x \in E$ there is a uniquely determined sequence of numbers $[\xi_r, R]$ such that

$$x = \lim_s \sum_{r=0}^{s} \xi_r e_r .$$

For each basis the correspondence $x \to \xi_s$ defines linear forms $f_s$ on $E$ with $\xi_s = \langle x, f_s \rangle$. Here we have

$$\langle e_r, f_s \rangle = \delta_{rs} \quad \text{for } r, s \in R .$$

**10.1.2.** We say that a basis $[e_r, R]$ is **equicontinuous** if for each zero neighborhood $U \in \mathfrak{U}(E)$ there is a zero neighborhood $V \in \mathfrak{U}(E)$ such that the inequalities

$$|\langle x, f_r \rangle| \, p_U(e_r) \leqq p_V(x) \quad \text{for } x \in E \text{ and } r \in R$$

are valid. In particular, all linear forms $f_r$ are continuous.

Without proof we mention the following

**Theorem.** *Each basis in an (F)-space is equicontinuous.*

**10.1.3.** We call an equicontinuous basis $[e_r, R]$ **absolute** if for each zero neighborhood $U \in \mathfrak{U}(E)$ there is a zero neighborhood $V \in \mathfrak{U}(E)$ for which the inequalities

$$\sum_R |\langle x, f_r \rangle| \, p_U(e_r) \leqq p_V(x) \quad \text{for } x \in E$$

hold. Then for all elements $x \in E$ we have the identity

$$x = \sum_R \langle x, f_r \rangle \, e_r ,$$

where the series on the right hand side is absolutely summable.

**10.1.4.** Complete locally convex spaces in which there is an absolute equicontinuous basis can be very easily represented. In fact we have the

**Theorem.** *Each complete locally convex space $E$ with an absolute equicontinuous basis $[e_r, R]$ can be identified with a sequence space $\Lambda$.*

*Proof.* We set

$$P = \{[p_U(e_r), R] : U \in \mathfrak{U}(E)\}$$

and construct the associated sequence space $\Lambda$, whose locally convex topology is obtained from the semi-norms

$$q_U[\xi_r, R] = \sum_R |\xi_r| \, p_U(e_r) \quad \text{with } U \in \mathfrak{U}(E) .$$

Since, by hypothesis, there is for each zero neighborhood $U \in \mathfrak{U}(E)$ a zero neighborhood $V \in \mathfrak{U}(E)$ with

$$q_U[\langle x, f_r\rangle, R] = \sum_R |\langle x, f_r\rangle| \, p_U(x) \leqq p_V(x) \quad \text{for } x \in E\,,$$

the equation

$$K x = [\langle x, f_r\rangle, R]$$

defines a one-to-one continuous linear mapping from $E$ into $\Lambda$. Since all families $[\xi_r e_r, R]$ with $[\xi_r, R] \in \Lambda$ in $E$ are absolutely summable, we can set

$$x = \sum_R \xi_r e_r.$$

But then the relation $K x = [\xi_r, R]$ is valid and we have shown that $K$ is also a mapping onto $\Lambda$. Finally, the continuity of the inverse mapping $K^{-1}$ follows from the inequality

$$p_U\left(\sum_R \xi_r e_r\right) \leqq \sum_R |\xi_r| \, p_U(e_r) = q_U[\xi_r, R]\,,$$

which is valid for $[\xi_r, R] \in \Lambda$ and $U \in \mathfrak{U}(E)$.

**10.1.5.** Since the unit vectors $e_s = [\delta_r^{[s]}, R]$ form an absolute equicontinuous basis in every sequence space $\Lambda$, we get from 10.1.4 the

**Theorem.** *The collection of all complete locally convex spaces in which there is an absolute equicontinuous basis coincides with the collection of all sequence spaces.*

## 10.2. Representation of Nuclear Locally Convex Spaces with Basis

**10.2.1.** Of greatest significance for the theory of nuclear locally convex spaces is the

**Basis Theorem.** *Each equicontinuous basis $[e_r, R]$ in a nuclear locally convex space $E$ is absolute.*

*Proof.* By hypothesis, there is for each zero neighborhood $U \in \mathfrak{U}(E)$ a zero neighborhood $U_0 \in \mathfrak{U}(E)$ with

$$|\langle y, f_r\rangle| \, p_U(e_r) \leqq p_{U_0}(y) \quad \text{for } y \in E \text{ and } r \in R.$$

Therefore, for each set $\mathfrak{r} \in \mathfrak{F}(R)$ we obtain a continuous linear mapping of $E(U_0)$ into $\boldsymbol{m}_{\mathfrak{r}}$ by the formula

$$S[y(U_0)] = [\langle y, f_r\rangle \, p_U(e_r), \mathfrak{r}]$$

and $\beta(S) \leqq 1$.

Repeating the first step we determine for a zero neighborhood $V_0 \in \mathfrak{U}(E)$ a zero neighborhood $V \in \mathfrak{U}(E)$ with

$$|\langle x, f_r\rangle| \, p_{V_0}(e_r) \leqq p_V(x) \quad \text{for } x \in E \text{ and } r \in R.$$

For each family $[\alpha_r, R] \in \boldsymbol{l}_R^2$ with $\boldsymbol{\lambda}_2[\alpha_r, R] \leq 1$ and all elements $x \in V$ a continuous linear mapping $T$ from $\boldsymbol{m}_\mathfrak{r}$ into $E(V_0)$ is defined by the formula

$$T[\xi_r, \mathfrak{r}] = \sum_{\mathfrak{r}} \alpha_r^2 \langle x, f_r \rangle\, \xi_r\, e_r(V_0)\,.$$

Since for each family $[\xi_r, \mathfrak{r}]$ with $|\xi_r| \leq 1$, we have the inequality

$$p(T[\xi_r, \mathfrak{r}]) \leq \sum_{\mathfrak{r}} |\alpha_r|^2\, |\langle x, f_r \rangle|\, p_{V_0}(e_r) \leq 1\,,$$

and it follows that $\boldsymbol{\beta}(T) \leq 1$.

Since the locally convex space $E$ is assumed to be nuclear we can by Theorem 8.6.1 choose the zero neighborhood $V_0$ in such a way that the canonical mapping from $E(V_0)$ onto $E(U_0)$ is of type $\boldsymbol{l}^{1/2}$. We can also assume that $\boldsymbol{\varrho}_{1/2}(E(V_0, U_0)) \leq 1$. Therefore, the estimate

$$\boldsymbol{\varrho}_{1/2}(K) \leq \boldsymbol{\beta}(S)\, \boldsymbol{\varrho}_{1/2}(E(V_0, U_0))\, \boldsymbol{\beta}(T) \leq 1$$

for the mapping $K = S\, E(V_0, U_0)\, T$ follows from Proposition 8.2.8.

But since $K \in \mathfrak{L}(\boldsymbol{m}_\mathfrak{r}, \boldsymbol{m}_\mathfrak{r})$ has the form

$$K[\xi_r, \mathfrak{r}] = [\alpha_r^2 \langle x, f_r \rangle\, p_U(e_r)\, \xi_r, \mathfrak{r}]\,,$$

the inequality

$$\sum_{\mathfrak{r}} |\alpha_r|\, |\langle x, f_r \rangle\, p_U(e_r)|^{1/2} \leq \boldsymbol{\varrho}_{1/2}(K) \leq 1$$

must hold by Proposition 8.2.9, and from it we obtain

$$\sum_{\mathfrak{r}} |\langle x, f_r \rangle|\, p_U(e_r) \leq 1$$

on the basis of Lemma 1.1.7. But since $\mathfrak{r}$ was an arbitrary set in $\mathfrak{F}(R)$, we get

$$\sum_R |\langle x, f_r \rangle|\, p_U(e_r) \leq 1 \quad \text{for all} \quad x \in V\,.$$

Consequently,

$$\sum_R |\langle x, f_r \rangle|\, p_U(e_r) \leq p_V(x) \quad \text{with} \quad x \in E$$

and our assertion is proved.

**10.2.2.** By combining the results in 10.1.5 and 10.2.1 we obtain the

**Theorem.** *The collection of all nuclear complete locally convex spaces in which there is an equicontinuous basis coincides with the collection of all nuclear sequence spaces.*

**10.2.3.** On the basis of the previous theorem all investigations of nuclear complete locally convex spaces with an equicontinuous basis reduce to the study of nuclear sequence spaces. In particular, this is true for the determination of wether two such spaces are isomorphic.

It is clear that two sequence spaces are isomorphic if they can be transformed into another by a **permutation** of $R$ and a **diagonal transformation**

$$D[\xi_r, R] = [\delta_r \xi_r, R] \quad \text{with} \quad \delta_r \neq 0 .$$

Whether all sequence spaces isomorphic to a nuclear sequence space can be obtained in this way is not yet known. Nevertheless, it can be proved that this statement is correct for nuclear power series spaces (B. Mitiagin [4]).

## 10.3. Bases in Special Nuclear Locally Convex Spaces

**10.3.1.** The fundamental

**Problem.** *Is there a basis in every nuclear* $(F)$*-space?*
is still unsolved. In several interesting special cases, however, a positive answer has been determined.

**10.3.2.** A real linear space $E$ is called **semi-ordered** (vgl. H. H. Schaefer [1]), if a cone $P$ of positive elements is determined for which the following conditions are satisfied:

$(H_1)$ *$x, y \in P$ implies $x + y \in P$.*
$(H_2)$ *$x \in P$ and $\alpha \geqq 0$ implies $\alpha x \in P$.*
$(H_3)$ *$x \in P$ and $-x \in P$ implies $x = 0$.*

When $x - y \in P$ we write $x \geqq y$. If for each pair of elements $x$ and $y$, $\sup(x, y)$ always exists, then $E$ is designated as a **linear lattice** and we set $|x| = \sup(x, -x)$.

A **locally convex lattice** is a linear lattice with a locally convex topology which can be determined by a system of semi-norms $p$ with the property

(V) *$|x| \geqq |y|$ implies $p(x) \geqq p(y)$.*

A complete metric locally convex lattice is called an $(F)$-**lattice.** Y. Kōmura and S. Koshi [1] proved the important

**Theorem.** *Each nuclear* $(F)$*-lattice has a basis.*

More recent investigations by B. Walsh [1] have shown that nuclearity can be replaced by an essentially weaker property.

**10.3.3.** If $A$ is a positive definite self-adjoint operator in a Hilbert space $H$ those elements $x \in H$, for which the collection of powers $x, A x, A^2 x, \ldots$ are defined form a reflexive $(F)$-space $D(A^\infty)$, whose topology is determined by the norms

$$p_r(x) = (A^r x, x)^{1/2} .$$

A. Pietsch [10] has shown that this $(F)$-space is nuclear if and only if the operator $A$ has a pure point spectrum such that the inequality

$$\sum_R \lambda_r^{-\varrho} < +\infty$$

holds for some positive number $\varrho$. The eigenvalues are counted with respect to their algebraic multiplicity. The associated eigenvectors $e_r$ then form a basis in $D(A^\infty)$. This method can be applied to construct bases in many nuclear function spaces (see H. Triebel [1], [2]).

A related existence theorem for bases in nuclear $(F)$-spaces is due to W. Wojtyński [1].

**10.3.4.** In the nuclear locally convex space $\mathcal{S}$ the Hermitian functions

$$h_r(t) = e^{t^2/2} \frac{d^r}{dt^r} (e^{-t^2})$$

form a basis.

**10.3.5.** In the nuclear locally convex space $\mathcal{E}[-1, +1]$ the Chebyshev polynomials

$$T_r(t) = \cos(r \arccos t)$$

form a basis.

**10.3.6.** In the nuclear locally convex space $\mathcal{E}_0[-1, +1]$ the transformed Hermitian functions

$$g_r(t) = h_r\left(\operatorname{tg} \frac{\pi}{2} t\right)$$

form a basis since the correspondence

$$[f(t)] \to \left[f\left(\operatorname{tg} \frac{\pi}{2} t\right)\right]$$

is an isomorphism between the spaces $\mathcal{S}$ and $\mathcal{E}_0[-1, +1]$.

**10.3.7.** In the nuclear locally convex space of the harmonic functions on a 3-dimensional open ball, the spherical functions form a basis.

**10.3.8.** In the nuclear locally convex space of analytic functions on the open disc, the powers $z^r$ form a basis.

**10.3.9.** The nuclear locally convex spaces $\mathcal{E}[-1, +1]$, $\mathcal{E}_0[-1, +1]$ and $\mathcal{S}$ can all be identified with the nuclear sequence space $\Sigma$ by the transformation defined in 10.1.4. They are therefore isomorphic to one another. Further isomorphism theorems are found in the work of Z. Ogrodzka [1], S. Rolewicz [3] and H. Triebel [1], [2].

Chapter 11

# Universal Nuclear Locally Convex Spaces

It is well known that every normed space can be embedded in a Banach space $\mathscr{C}(M)$. Here $M = [0, 1]$ can be used for separable normed spaces. A. Grothendieck conjectured that in the theory of nuclear locally convex spaces, the $I$-fold topological product $(\Sigma)^I$ of the sequence space $\Sigma$ takes the place of the Banach space $\mathscr{C}(M)$. This important structure theorem was recently proved by T. and Y. Kōmura. For nuclear $(F)$-spaces with basis this assertion had previously been proven by C. Bessaga and A. Pełczyński.

As a simple consequence we have the proposition that each nuclear $(F)$-space can be identified with a linear subspace of the function space $\mathscr{E}$.

A further universality theorem due to A. Martineau asserts that each $s$-nuclear locally convex space can be embedded in an $I$-fold topological product $(\Sigma')^I$ of the dual sequence space $\Sigma'$.

## 11.1. Embedding in the Product Space $(\Sigma)^I$

**11.1.1.** We have the

**Theorem.** *Each nuclear locally convex space $E$ can be identified with a linear subspace of $(\Sigma)^I$.*

*Proof.* We consider in $E$ an arbitrary fundamental system

$$\mathfrak{U}_{\mathfrak{F}}(E) = \{U_i : i \in I\}$$

of zero neighborhoods. On the basis of a dual analog of Theorem 9.5.3, there is for each zero neighborhood $U_i$ a sequence of linear forms $a_r^{[i]}$ with

$$U_i^0 \subset A\,[a_r^{[i]}, R]$$

such that the sets

$$G_n^{[i]} = \{(r+1)^n\, a_r^{[i]} : r \in R\}$$

are equicontinuous for all natural numbers $n$. Consequently, we get

$$p_i(x) = \sup\{|\langle x, a\rangle| : a \in U_i^0\} \leq \sup\{|\langle x, a_r^{[i]}\rangle| : r \in R\}$$

and

$$q_n[\langle x, a_r^{[i]}\rangle, R] = \sup\{|\langle x, a_r^{[i]}\rangle|\,(r+1)^n : r \in R\} \leq \sup\{|\langle x, a\rangle| : a \in G_n^{[i]}\}.$$

Since by Theorem 6.1.3 we can determine the locally convex topology of $\Sigma$ with the semi-norms

$$q_n[\xi_r, R] = \sup\{|\xi_r| (r+1)^n : r \in R\},$$

the formula

$$K_i x = [\langle x, a_r^{[i]}\rangle, R]$$

defines continuous linear mappings $K_i$ from $E$ into $\Sigma$, from which we get the desired embedding $K$ by setting

$$K x = [K_i x, I].$$

**11.1.2.** It was shown by B. Mitiagin [4] that the locally convex space $\mathscr{E}$ is isomorphic to the $N$-fold topological product $(\Sigma)^N$. We shall here prove a weaker statement.

**Proposition.** *The nuclear locally convex space $(\Sigma)^N$ can be identified with a linear subspace of $\mathscr{E}$.*

*Proof.* The topological product

$$\Pi\, \mathscr{E}_0 [2(n-1)\pi, 2n\pi]$$

can be considered as a linear subspace of $\mathscr{E}$ if we associate with the family $[f_n, N]$ with $f_n \in \mathscr{E}_0[2(n-1)\pi, 2n\pi]$ the function $f \in \mathscr{E}$ with

$$f(t) = 0 \quad \text{for } t \leqq 0 \quad \text{and} \quad f(t) = f_n(t) \quad \text{for } 2(n-1)\pi \leqq t \leqq 2n\pi.$$

But since all locally convex spaces $\mathscr{E}_0[2(n-1)\pi, 2n\pi]$ are isomorphic to the sequence space $\Sigma$ by 10.3.9, we have found the desired embedding of $(\Sigma)^N$ into $\mathscr{E}$.

**11.1.3.** By combining the results of 11.1.1 and 11.1.2 we obtain the

**Theorem.** *Each nuclear (F)-space can be identified with a linear subspace of $(\Sigma)^N$ or $\mathscr{E}$.*

**11.1.4.** The preceeding considerations have shown that the spaces $(\Sigma)^I$ and $\mathscr{E}$ occupy the same position among nuclear spaces that the space $\mathscr{C}(M)$ has among normed spaces. Therefore we have the following

**Problem.** *Can the nuclear spaces $(\Sigma)^I$ and $\mathscr{E}$ be characterized abstractly?*

**11.1.5.** Since each arbitrary Banach space can be represented as the quotient space of a special Banach space $\boldsymbol{l}_I^1$, the question arises whether an analogous statement holds in the theory of nuclear locally convex spaces. A. Martineau [1] has formulated in this regard the following

**Problem.** *Can each nuclear (F)-space be represented as the quotient space of the sequence space $\Sigma$?*

## 11.2. Embedding in the Product Space $(\Sigma')^I$

**11.2.1.** A locally convex space $E$ is called **s-nuclear** if it contains a fundamental system $\mathfrak{U}_{\mathfrak{F}}(F)$ of zero neighborhoods which has the following two equivalent properties:

$(S')$ *For each zero neighborhood* $U \in \mathfrak{U}_{\mathfrak{F}}(E)$ *there exists a zero neighborhood* $V \in \mathfrak{U}_{\mathfrak{F}}(E)$ *with* $V < U$, *such that the canonical mapping from* $E(V)$ *onto* $E(U)$ *is of type* **s**.

$(S)$ *For each zero neighborhood* $U \in \mathfrak{U}_{\mathfrak{F}}(E)$ *there exists a zero neighborhood* $V \in \mathfrak{U}_{\mathfrak{F}}(E)$ *with* $V < U$, *such that the canonical mapping from* $E'(U^0)$ *into* $E'(V^0)$ *is of type* **s**.

**11.2.2.** Obviously every *s*-nuclear locally convex space is nuclear. On the other hand, there are many nuclear locally convex spaces which are not *s*-nuclear. As an example we mention the sequence space $\Sigma$.

The statement formulated in 8.5.2 leads directly to the following

**Theorem.** *Each nuclear dual metric locally convex space is also s-nuclear.*

**11.2.3.** Since *s*-nuclear spaces have the same permanence properties as nuclear spaces (cf.. B. S. Brudovskij [1] and A. Martineau [1]), all $I$-fold topological products $(\Sigma')^I$ of the *s*-nuclear space $\Sigma'$ are *s*-nuclear as well as the linear subspaces of these products. On the other hand, A. Martineau [1] proved the following

**Theorem.** *Each s-nuclear locally convex space* $E$ *can be identified with a linear subspace of* $(\Sigma')^I$.

*Proof.* We consider in $E$ an arbitrary fundamental system

$$\mathfrak{U}_{\mathfrak{F}}(E) = \{U_i : i \in I\}$$

of zero neighborhoods and determine further zero neighborhoods $V_i$, such that the canonical mappings of $E(V_i)$ onto $E(U_i)$ are of type **s**. By Theorem 8.5.6 there are linear forms $a_r^{[i]} \in V_i^0$, elements $y_r^{[i]} \in U_i$ and rapidly decreasing sequences of numbers $[\lambda_r^{[i]}, R]$ with

$$x(U_i) = \sum_R \lambda_r^{[i]} \langle x, a_r^{[i]}\rangle \, y_r^{[i]}(U_i) \quad \text{for } x \in E .$$

Consequently, we have

$$p_i(x) = \sup\{|\langle x, a\rangle| : a \in U_i^0\} \leqq \sum_R |\lambda_r^{[i]}| \, |\langle x, a_r^{[i]}\rangle|$$

and

$$p[\langle x, a_r^{[i]}\rangle, R] = \sum_R |\langle x, a_r^{[i]}\rangle| \, \varrho_r \leqq \sum_R \varrho_r \sup\{|\langle x, a\rangle| : a \in V_i^0\} .$$

Since we can determine the locally convex topology of $\Sigma'$ from the semi-norms

$$p[\xi_r, R] = \sum_R |\xi_r| \varrho \quad \text{with} \quad [\varrho_r, R] \in \Sigma \quad \text{and} \quad \varrho_r \geqq 0\,,$$

the equation

$$K_i\, x = [\langle x, a_r^{[i]} \rangle, R]$$

defines continuous linear mappings $K_i$ from $E$ into $\Sigma'$, and we get the desired embedding

$$K\, x = [K_i\, x, I]$$

by taking all of the $K_i$ together.

# Bibliography


Ahiezer, N. I., Glazman, I. M:

1. Theory of linear Operators in Hilbert Space, Moscow 1950.

Amemiya, I., Shiga, K.:

1. On tensor products of Banach spaces, Kodai Math. Sem. Report **9**, 161–178 (1957).

Banach, S.:

1. Théorie des opérations linéaires, Warsaw 1932.

Bauer, H.:

1. Harmonische Räume und ihre Potentialtheorie, Lecture Notes in Mathematics **22**, Berlin/Heidelberg/New York: Springer 1966.

Bessaga, C., Pełczyński, A.:

1. On the embedding of nuclear spaces into the space of infinitely differentiable functions on the line, DAN U.S.S.R. **134**, 745–748 (1966).

Bessaga, C., Pełczyński, A., Rolewicz, S.:

1. Approximative dimension of linear topological spaces and some of its applications, Reports of the Conference on Functional Analysis, Warsaw 1960; Studia Math. Seria specjalna **1**, 27–29 (1963).
2. On diametral approximative dimension and linear homogeneity of $F$-spaces, Bull. Acad. Polon. Sci. **9**, 677–683 (1961).
3. Some remarks on Dragilev's type theorem, Studia Math. **31**, 307–318 (1968).

Bogdanowicz, W.:

1. A proof of Schwartz's theorem on kernels, Studia Math. **20**, 77–85 (1961).

Bourbaki, N.:

1. Topologie générale, Paris 1940/49.
2. Algébre multilinéaire, Paris 1948.
3. Intégration, Paris 1952.
4. Espaces vectoriels topologiques, Paris 1953/55.

Brudovskij, B. S.:

1. The associated nuclear topology, mappings of type $s$ and strongly nuclear spaces, DAN U.S.S.R. **178**, 271–273 (1968).
2. $s$-type mappings of locally convex spaces, DAN U.S.S.R. **180**, 15–17 (1968).

Chang, S. H.:

1. On the distribution of the characteristic values and singular values of linear integral equations, Trans. Amer. Math. Soc. **67**, 351–367 (1949).

Chevet, S.:

1. Sur certains produits tensoriels topologiques d'espaces de Banach, C. R. Acad. Sci. Paris **266**, 413–415 (1968).

Day, M. M.:

1. Normed linear spaces, Berlin/Göttingen/Heidelberg 1958.

Dollinger, M. B.:

1. Nuclear topologies consistent with a duality, Proc. Amer. Math. Soc. **23**, 565–568 (1969).

Dragilev, M. M.:

1. The canonical form of the basis in the space of analytic functions, Usp. Mat. Nauk **15** (2), 181—188 (1960).
2. On regular bases in nuclear spaces, Mat. Sb. **68**, 153—173 (1965).

Dragilev, M. M., Kondakov, V. P.:

1. On a class of nuclear locally convex spaces, Mat. Zametki **8**, 169—179 (1970).

Dubinski, E.:

1. Equivalent nuclear systems, Studia Math. **38**, 374—379 (1970).

Dubinski, E., Ramanujan, M. S.:

1. On $s$-nuclearity.

Dunford, N., Schwartz, J. T.:

1. Linear Operators, New York 1958.

Dvoretzky, A., Rogers, C. A.:

1. Absolute and unconditional convergence in normed linear spaces, Proc. Nat. Acad. Sci. USA **36**, 192—197 (1950).

Dynin, A. S.:

1. On various kind of nuclear spaces, DAN U.S.S.R. **121**, 790—792 (1958).
2. Bases in nuclear spaces, Reports of the Conference on Functional Analysis, Warsaw 1960.

Dynin, A. S., Mitiagin, B. S.:

1. Criterion for nuclearity in terms of approximative dimension, Bull. Acad. Polon. Sci. **8**, 535—540 (1960).

Ehrenpreis, L.:

1. On the theorem of kernels of Schwartz, Proc. Amer. Math. Soc. **7**, 713—718 (1956).

Eizenberg, L. A.:

1. On spaces of analytic functions on the $(p, q)$-circular domain, DAN U.S.S.R **136**, 521—524 (1961).

Eizenberg, L. A., Mitiagin, B. S.:

1. The space of analytic functions on a multicircular domain, Sib. Mat. J. **1**, 153—170 (1960).

Erochin, W. D.:

1. On the asymptotic of $\varepsilon$-entropy of analytic functions, DAN U.S.S.R. **120**, 949—952 (1958).

Fiedler, M., Pták, V.:

1. Sur la meilleure approximation des transformations linéaires par des transformations de rang prescript, C. R. Acad. Sci. Paris **254**, 3805—3807 (1962).

Floret, K., Wloka, J.:

1. Einführung in die Theorie der lokalkonvexen Räume, Lecture Notes in Mathematics **56**, Berlin/Heidelberg/New York: Springer 1968.

Gask, H.:

1. A proof of Schwartz's kernel theorem, Math. Scand. **8**, 327—332 (1960).

Gelfand, I. M., Šilov, G. E.:

1. Generalized Functions, Volumes 1, 2, Moscow 1958.

Gelfand, I. M., Vilenkin, N. J.:

1. Generalized Functions, Volume 4, Moscow 1961.

Gelfand, I. M.:

1. On some problems of functional analysis, Usp. Mat. Nauk. **11** (6), 3—12 (1956).

Grothendieck, A.:

1. Sur une notion de produit tensoriel topologique d'espaces vectoriels topologiques, et une classe remarquable d'espaces vectoriels liées à cette notion, C. R. Acad. Sci. Paris **233**, 1556—1558 (1951).
2. Résumé des résultats essentiels dans la théorie des produits tensoriels topologiques et des espaces nucléaires, Ann. Inst. Fourier **4**, 73—112 (1954).
3. Produits tensoriels topologiques et espaces nucléaires, Memoires Amer. Math. Soc. **16** (1955).
4. Resultats noveaux dans la théorie des operations linéaires, C. R. Acad. Sci. Paris **239**, 577—579, 607—609 (1954).
5. Résumé de la théorie métrique des produits tensoriels topologiques, Boletim Soc. Mat. Sao Paulo **8**, 1—79 (1956).
6. Sur certains classes de suites dans les espaces de Banach, et le théoréme de Dvoretzky-Rogers, Boletim Soc. Mat. Sao Paulo **8**, 81—110 (1956).
7. Sur certains espaces de fonctions holomorph, J. reine angew. Math. **192**, 35—64, 77—95 (1953).
8. Espaces vectoriels topologiques, Sao Paulo 1954.

Halmos, P. R.:

1. Measure theory, Princeton 1950.
2. Introduction to Hilbert spaces and the theory of spectral multiplicity, New York 1951.

Hilbert, D.:

1. Grundzüge einer allgemeinen Theorie der linearen Integralgleichungen, Leipzig 1912.

Hildebrandt, T. H.:

1. On unconditional convergence in normed vector spaces, Bull. Amer. Math. Soc. **46**, 959—962 (1940).

Hinrichsen, D.:

1. Randintegrale und nukleare Funktionenräume, Ann. Inst. Fourier **17**, 225 bis 271 (1967).

Hogbe-Nlend, H.:

1. Complétion, tenseurs et nucléarité en bornologie, J. Math. Pures Appl. **49**, 193—288 (1970).
2. Une caractérisation intrinsèque du dual forte d'un espace de Fréchet nucléaire ou ultra-nucléaire, C.R. Acad. Sci. Paris **272** A, 244—246 (1971).
3. Théorie des Bornologies et Applications, Lecture Notes in Mathematics **213**, Berlin/Heidelberg/New York: Springer 1971.

Horn, A.:

1. On the singular values of a product of completely continuous operators, Proc. Nat. Acad. Sci. USA **36**, 374—375 (1950).

Kaczmarz, S., Steinhaus, H.:

1. Theorie der Orthogonalreihen, Warschau 1935.

Kamil, J.:

1. Zwei Charakterisierungen der nuklearen lokalkonvexen Räume, Comment. Math. Univ. Carolinae **8**, 117—128 (1967).

Kantorovič, L. W., Akilov, G. P.:

Functional analysis in normed spaces, Moscow 1959.

Kelley, J. L.:

1. General topology, Princeton 1955.

Kelley, J. L., Namioka, I.:

1. Linear topological spaces, Princeton 1963.

Köthe, G., Toeplitz, O.:
1. Lineare Räume mit unendlich vielen Koordinaten und Ringe unendlicher Matrizen, J. reine angew. Math. **171**, 193—226 (1934).

Köthe, G.:
1. Die Stufenräume, eine einfache Klasse linearer vollkommener Räume, Math. Z. **51**, 317—345 (1948).
2. Neubegründung der Theorie der vollkommenen Räume, Math. Nachr. **4**, 70—80 (1951).
3. Dualität in der Funktionentheorie, J. reine angew. Math. **191**, 30—49 (1953).
4. Topologische lineare Räume, Berlin/Göttingen/Heidelberg: Springer 1960.
5. Über nukleare Folgenräume, Studia Math. **31**, 267—271 (1968).
6. Stark nukleare Folgenräume, J. Fac. Sci, Univ. Tokyo Sect. I, **17**, 291—296 (1970).

Kolmogorov, A. N.:
1. On some asymptotic characterizations of totally bounded metric spaces, DAN U.S.S.R., **108**, 385—388 (1956).
2. On the linear dimension of topological vector spaces, DAN U.S.S.R. **120**, 239—241 (1958).

Kolmogorov, A. N., Tihomirov, W. M.:
1. The $\varepsilon$-entropy and $\varepsilon$-capacity of sets in function spaces, Usp. Mat. Nauk. **14** (2), 3—86 (1959).

Kōmura, Y.:
1. Die Nuklearität der Lösungsräume der Hypoelliptischen Gleichungen, Funkcial. Ekvac. **9**, 313—324 (1966).

Kōmura, T., Kōmura, Y.:
1. Über die Einbettung der nuklearen Räume in $(s)^A$, Math. Ann. **162**, 284—288 (1966).

Kōmura, Y., Koshi, S.:
1. Nuclear vector lattices, Math. Ann. **163**, 105—110 (1966).

Kostjušenko, A. G., Mitiagin, B. S.:
1. Positive definite functions on nuclear spaces, Trudy. Mosc. Mat. Ob-va. **9**, 283—316 (1960).

Krein, M. G., Krasnoselski, M. A., Milman, D. P.:
1. On defect numbers of operators in Banach spaces and some geometrical questians, Sb. trudov in-ta matem. AN. U.S.S.R. **11**, 97—112 (1948).

Kwapień, S.:
1. On a theorem of L. Schwartz and its application to aboslutely summing operators, Studia Math. **38**, 193—201 (1970).

Lindenstrauss, J., Pełczyński, A.:
1. Absolutely summing operators between $L_p$-spaces, Studia Math. **29**, 275—326 (1968).

Littlewood, J. E.:
1. On bounded bilinear forms in an infinity of variables, Quart, J. Math. (Oxford) **1**, 64—174 (1930).

Loeb, P. A., Walsh, B.:
1. Nuclearity in axiomatic potential theory, Bull. Amer. Math. Soc. **72**, 685—689 (1966).

Mac Phail, M. S.:
1. Absolute and unconditional convergence, Bull. Amer. Math. Soc. **53**, 121—123 (1947).

Martineau, A.:
1. Sur une proprieté universelle de l'espace des distribution de M. Schwartz, C. R. Acad. Sci. Paris **259**, 3162—3164 (1964).

Maurin, K.:
1. Abbildungen vom Hilbert-Schmidtschen Typus und ihre Anwendungen, Math. Scand. **9**, 359—371 (1961).

Minlos, R. A.:
1. The extension of a generalized stochastic process to additive measures, DAN U.S.S.R. **119**, 439—442 (1948).
2. The generalized stochastic processes and their extension to measures, Trudy Mosc. Mat. Ob-va. **8**, 497—518 (1959).

Mitiagin, B. S.:
1. Nuclearity and other properties of spaces of type *S*, Trudy Mosc. Mat. Ob-va **9**, 377—428 (1960).
2. The connection between $\varepsilon$-entropy, approximation and nuclearity of compact sets in a linear space, DAN U.S.S.R. **134**, 765—768 (1960).
3. Nuclear Riesz scales, DAN U.S.S.R. **137**, 519—522 (1961).
4. Approximative dimensions and bases in nuclear spaces, Usp. Math. Nauk **16** (4), 73—132 (1961).
5. A remark on quasi-invariant measure, Usp. Mat. Nauk **16** (5), 191—193 (1961).
6. Fréchet spaces with a unique unconditional basis, Studia Math. **38**, 23—34 (1970).

Moore, E. H.:
1. General analysis II, Memoirs Amer. Phil. Soc. Philadelphia 1939.

Munroe, M. E.:
1. Absolute and unconditional convergence in Banach spaces, Duke Math. J. **13**, 351—365 (1946).

Murray, F. J., Neumann, J. von:
1. On rings of operators Ann. of Math. **37**, 116—229 (1936).

Neumann, J. von:
1. Some matrix inequalities and metrization of matrix-spaces, Izv. in-ta matem. i mech. Tomsk. Univ. **1**, 286—300 (1937).

Ogradzka, Z.:
1. On simultaneous extension of infinitely differentiable functions, Studia Math. **28**, 193—207 (1967).

Orlicz, W.:
1. Beiträge zur Theorie der Orthogonalreihen II, Studia Math. **1**, 241—255 (1929).
2. Über unbedingte Konvergenz in Funktionenräumen, Studia Math. **4**, 33—37, 41—47 (1933).

Pełczyński, A.:
1. On the approximation of S-spaces by finite dimensional spaces, Bull. Acad. Polon. Sci. **5**, 879—881 (1957).
2. Proof of a theorem of Grothendieck on the characterization of nuclear spaces, Prace Mat. **7**, 155—167 (1962).
3. A characterization of Hilbert-Schmidt operators, Studia Math. **28**, 355—360 (1967).

Pełczyński, A., Szlenk, W.:
1. Sur l'injection naturelle de l'espace $(l)$ dans l'espaces $(l_p)$, Coll. Math. **10**, 313—323 (1963).

Persson, A., Pietsch, A.:
1. $p$-nukleare und $p$-integrale Abbildungen in Banachräumen, Studia Math. **33**, 19—62 (1969).

Pettis, B. J.:
1. On integration in vector spaces, Trans, Amer. Math. Soc. **44**, 277—304 (1938).

Pietsch, A.:
1. Unbedingte und absolute Summierbarkeit in $F$-Räumen, Math. Nachr. **23**, 215—222 (1961).
2. Eine neue Charakterisierung der nuklearen lokalkonvexen Räume, Monatsberichte Dt. Akad. Wiss. **4**, 325—329 (1962).
3. Eine neue Charakterisierung der nuklearen lokalkonvexen Räume, Math. Nachr. **25**, 31—36, 51—58 (1963).
4. Absolut summierende Abbildungen in lokalkonvexen Räumen, Math. Nachr. **27**, 77—103 (1963).
5. Einige neue Klassen von kompakten linearen Abbildungen, Revue de Math. pures et appl. (Bukarest) **8**, 427—447 (1963).
6. Zur Theorie der topologischen Tensorprodukte, Math. Nach. **25**, 19—31 (1963).
7. Verallgemeinerte vollkommene Folgenräume, Berlin 1962.
8. Quasinukleare Abbildungen in normierten Räumen, Math. Ann. **165**, 76—90 (1966).
9. Absolute Summierbarkeit in Vektorverbänden, Math. Nachr. **26**, 15—23 (1963).
10. Über die Erzeugung von $(F)$-Räumen durch selbstadjundierte Operatoren, Math. Ann. **164**, 219—224 (1966).
11. Nukleare Funktionenräume, Math. Nachr. **33**, 377—384 (1967).
12. $(F)$-Räume mit absoluter Basis, Studia Math. **26**, 233—238 (1966).
13. Absolut $p$-summierende Abbildungen in normierten Räumen, Studia Math. **28**, 333—353 (1967).
14. Hilbert-Schmidt-Abbildungen in Banach-Räumen, Math. Nachr. **37**, 237 bis 245 (1968).
15. Ideale von Operatoren in Banachräumen. Mitteilungen der MGdDDR 1968, S. 1—13.
16. Ideale von $S_p$-Operatoren in Banachräumen, Studia Math. **38**, 59—69 (1970).
17. $l_p$-faktorisierbare Operatoren in Banachräumen, Acta Sci, Math. (Szeged) **31**, 117—123 (1970).
18. Absolutely-$p$-summing operators in $L_r$-spaces I, II, Sem. Goulaouic-Schwartz, Paris 1970/71.
19. Ideals of operators on Banach spaces and nuclear locally convex spaces, Proc. III. Symp. General Topology, Prag 1971.

Pietsch, A., Triebel, H.:
1. Interpolationstheorie für Banachideale von beschränkten linearen Operatoren, Studia Math. **31**, 203—217 (1968).

Rham, G. de:
1. Variétés différentiables, Paris 1955.

Raikov, D. A.:

1. On a property of nuclear spaces, Usp. Mat. Nauk **12** (5), 231–236 (1957).

Ramanujan, M. S.:

1. Power series spaces $\Lambda(P)$ and associated $(\Lambda)$-nuclearity, Math. Ann. **189**, 161–168 (1970).

Rolewicz, S.:

1. Remarks on linear metric Montel spaces, Bull. Acad. Polon. Sci. **7**, 195–197 (1959).
2. On the isomorphism and approximative dimension of spaces of analytic functions, DAN U.S.S.R. **133**, 32–33 (1960).
3. On spaces of holomorphic functions, Studia Math. **21**, 135–160 (1962).

Rosenberger, B.:

1. *s*-nukleare Räume, Thesis, Bonn 1970, Math. Nachr.
2. F-Normideale von Operatoren in normierten Räumen, BMBW-GMD-44 (1971).

Ruston, A. F.:

1. On the Fredholm theory of integral equations for operators belonging to the trace class of a general Banach space, Proc. London Math. Soc. (2) **53**, 109–124 (1951).
2. Direct products of Banach spaces and linear functional equations, Proc. London Math. Soc. (3), **1**, 327–384 (1951).

Rutovitz, D.:

1. Absolute and unconditional convergence in normed linear spaces, Proc. Cambridge Phil. Soc. **58**, 575–579 (1962).

Saphar, P.:

1. Applications à puissance nucléaire et applications de Hilbert-Schmidt dans le espaces de Banach, Ann. l'école norm. sup. 3. série **83**, 113–151 (1966).
2. Produits tensoriels topologiques et classes d'applications linéaires, C. R. Acad. Sci. Paris **266**, 526–528 (1968).
3. Comparaison de normes sur des produits tensoriels d'espaces de Banach. Applications, C. R. Acad. Sci. Paris **266**, 809–811 (1968).

Sazonov, W.:

1. A remark on characteristic functionals, Teor. Verojatnost. i Primenen **3**, 201–205 (1958).

Schaefer, H. H.:

1. Topological vectors spaces, New York 1965.

Schatten, R.:

1. On the direct product of Banach spaces, Trans. Amer. Math. Soc. **53**, 516–541 (1943).
2. The cross-space of linear transformations I, Ann. of Math. **47**, 73–84 (1946).
3. A theory of cross-spaces, Princeton 1950.
4. Norm ideals of completely continuous operators, Berlin/Göttingen/Heidelberg: Springer 1960.

Schatten, R., Neumann, J. von:

1. The cross-space of linear transformations II, III, Ann. of Math. **47**, 608–630 (1946) **49**, 557–582 (1948).

Schmidt, E.:

1. Zur Theorie der linearen und nicht linearen Integralgleichungen, Math. Ann. **63**, 433–476 (1907) **64**, 161–174 (1907).

Schock, E.:

1. Diametrale Dimension, approximative Dimension und Anwendungen, BMBW-GMD-43 (1971).

Schwartz, L.:
1. Théorie des noyaux, Proc. Intern. Congr. of Math. Cambridge Mass. Vol. **1**, 220—230 (1950).
2. Théorie des distributions, Paris 1950/51.
3. Espaces de fonctions différentiables à valeurs vectorielles, J. d'Analyse Math. **4**, 88—148 (1954/55).
4. Théorie des distributions à valeurs vectorielles, Ann. Inst. Fourier **7**, 1—141 (1957) **8**, 1—209 (1958).
5. (Séminaire) Produits tensoriels topologiques d'espaces vectoriels topologiques. Espaces vectoriels topologiques nucléaires. Applications, Paris 1953/54.
6. Mesures de Radon sur des espaces topologiques arbitraires, Paris 1964/65.
7. Applications radonifiantes, Sem., Paris 1969/70.

Singer, I.:
1. Sur les applications linéaires integrales des espaces de fonctions continues, Revue de Math. pures appl. (Bukarest) **4**, 391—401 (1959).

Spuhler, P.:
1. *s*-nukleare Räume, Thesis, Frankfurt 1970.

Sudakov, W. N.:
1. Linear sets with quasi-invariant measure. DAN U.S.S.R. **127**, 524—525 (1959).

Tanaka, S.:
1. $\varepsilon$-Entropie of subsets of the spaces of solutions of certain partial differential equations, J. of Math. Kyoto Univ. **6**, 313—322 (1967).

Terzioglu, T.:
1. Die diametrale Dimension von lokalkonvexen Räumen, Collectanea Math. **20**, 49—99 (1969).

Tihomirov, B. M.:
1. On the $n$-dimensional diameters of compact sets, DAN U.S.S.R. **130**, 734—737 (1960).
2. Diameters of sets in functional spaces and the theory of best approximation, Usp. Mat. Nauk. **15** (3), 81—120 (1960).

Tillmann, H. G.:
1. Dualität in der Potentialtheorie, Portugal. Math. **13**, 55—86 (1954).

Toeplitz O.:
1. Die linearen vollkommenen Räume der Funktionentheorie, Comm. math. Helvetici **23**, 222—224 (1949).

Treves, F.:
1. Topological vector spaces, distributions and kernels, New York—London 1967.

Triebel, H.:
1. Erzeugung nuklearer lokalkonvexer Räume durch singuläre Differentialoperatoren zweiter Ordnung, Math. Ann. **174**, 163—176 (1967).
2. Erzeugung des nuklearen lokalkonvexen Raumes $C^\infty(\overline{\Omega})$ durch elliptische Differentialoperatoren zweiter Ordnung, Math. Ann. **177**, 247—264 (1968).
3. Nukleare Funktionenräume und singuläre elliptische Differentialoperatoren, Studia Math. **38**, 285—311 (1970).

Umemura, Y.:
1. Measures on infinite dimensional vector spaces, Pub. Res. Inst. Math. Sci., Kyoto Univ. Ser. A, **1** (1965).

Waelbroeck, L.:
1. Espaces nucleaires, Sem., Brüssel 1969.
Walsh, B.:
1. On Characterizing Köthe Sequence Spaces as Vector Lattices, Math. Ann. **175**, 253—256 (1968).
2. Spaces of continuous functions characterized by kernel-like conditions, Math. Z. **109**, 71—86 (1969).
Weyl, H.:
1. Inequalities between the two kinds of eigenvalues of a linear transformation, Proc. Nat. Acad. Sci. USA **35**, 408—411 (1949).
Wilde, M. de:
1. Espaces de fonctions à valeurs dans un espace linéaire à semi-normes, Memoires Soc. R. Sc. de Liégé **12**, 2 (1966).
2. Sur les operateurs prènucléaires et integraux, Bull. Soc. R. de Liége **35**, 22—39 (1966).
Włoka, J.:
1. Kernel functions and nuclear spaces, Bull. Amer. Math. Soc. **71**, 720—723 (1965).
2. Reproduzierende Kerne und nukleare Räume, Math. Ann. **163**, 167—188 (1966)); **172**, 79—93 (1967).
3. Nukleare Räume aus M. K.-Funktionen, Math. Z. **92**, 295—306 (1966).
Wojtyński, W.:
1. On bases on certain countably hilbert spaces, Bull. Acad. Polon, Sci. **14**, 681—684 (1966).

# Index

approximation number, $r$-th (8.1.1) 121

Banach space (0.8.1) 11
basis (10.1.1) 172
—, absolute equicontinuous (10.1.3) 172
—, equicontinuous (10.1.2) 172
Basis Theorem (10.2.1) 173
bidual (0.6.9) 10
bilinear form, nuclear (7.4.2) 115
bipolar (0.6.4) 9

Cauchy-system (0.5.7) 8
—, bounded (0.5.7) 8
convergence of a directed system (0.1.5) 2

diameter, $r$-th (9.1.1) 144
— of $V$ with respect to $U$, $r$-th (9.2.1) 149
diagonal transformation (10.2.3) 175
dimension, approximative (9.7.2) 164
—, diametral (9.2.2) 149
—, dual approximative (9.7.7) 166
—, dual diametral (9.2.7) 151
dual, algebraic (0.3.5) 5
—, topological (0.6.1) 1
Duality Theorem (5.1.9) 89

elements, linearly independent (0.3.2) 4
—, orthogonal (0.9.4) 12
equivalence classes (0.3.4) 5
$\varepsilon$-content (9.6.1) 160
$\varepsilon$-content of $V$ with respect to $U$ (9.7.1) 164
$\varepsilon$-topology (1.2.3), (7.1.2) 23, 108

family, absolutely summable (1.4.1) 27
—, finite dimensional (1.6.1) 32
—, totally summable (1.5.1) 29
—, weakly summable (1.2.1) 29
— of numbers (1.1.1) 18
—, quadratically summable (1.1.7) 22
—, summable (1.1.1) 18

functions, characteristic (0.12.6) 17
—, Rademacher (2.4.1) 42
fundamental system of bounded subsets (0.5.5) 7
—, of neighborhoods (0.1.2) 1
—, of zero neighborhoods (0.5.1) 6

Hausdorff space (0.1.5) 2
—, compact (0.1.9) 3
Hilbert space (0.9.2) 12
Hilbert-Schmidt-mappings (2.5.1) 45
hull, closed (0.1.7) 2
—, complete (0.5.7) 8
—, locally convex (5.2.4) 92
—, quasi-complete (0.5.7) 8

inequality, Bessel's (0.9.4) 12
—, Hölder's (1.1.7) 22
—, Khinchin's (2.4.1) 43
isomorphism of locally convex spaces (9.2.3) 150

kernel, locally convex (5.2.3) 92
Kernel Theorem (7.4.3) 115

lattice, (F)- (10.3.2) 175
limit (0.1.5) 2
—, linear (10.3.2) 175
linear combination (0.3.2) 4
linear form (0.3.5) 5
locally convex space (0.5.1) 6

mapping, absolutely summing (2.1.1) 34
—, adjoint (0.10.2) 13
—, canonical (0.11.5) 16
—, compact (0.10.6) 14
—, continuous (0.1.11) 3
—, continuous linear (0.10.1) 13
—, dual (0.10.2) 13
—, finite (0.10.5) 14
—, Hilbert-Schmidt- (2.5.1) 45
—, linear (0.10.1) 13

mapping, nuclear (3.1.1) 49
—, precompact (0.10.6) 14
—, quasi-nuclear (3.2.3) 56
mapping of type $l^p$ (8.2.1) 125
mapping of type $s$ (8.5.1) 138
measure, Dirac (2.3.1) 39
—, positive Radon (0.12.2) 16
—, Radon (0.12.2) 16
—, of a subset (0.12.4) 17
metric (0.2.1) 3

neighborhood (0.1.2) 1
norm (0.8.1) 11

order of $V$ with respect to $U$ (9.8.1) 161
orthogonal (0.9.4) 12
orthonormal system (0.9.4) 12
—, complete (0.9.4) 12

partial sum (1.1.1), (1.3.6) 18, 26
polar (0.6.4) 9
power series space (6.1.5) 99
product, topological (0.1.10), (5.2.1) 3, 90
projection (0.10.7) 14
$\pi$-topology (1.4.2), (7.1.2) 27, 108
Property $(B)$ (1.5.5) 30
Property $(N)$ (4.1.1), (4.1.2) 64, 70
Property $(N')$ (4.1.1), (4.1.2) 69, 70
Property $(P)$ (4.1.5) 71
Property $(Q)$ (4.1.4) 71

quasi-norm (8.2.3) 125
quotient space (0.3.4) 5

range (0.10.1) 13
Riesz Representation Theorem (0.9.3) 12

scalar product (0.9.1) 11
semi-norm (0.4.1) 5
semi-scalar product (0.9.1) 71
sequence (0.1.5)
—, rapidly decreasing (9.5.3) 158
sequence of numbers, rapidly decreasing (8.5.5) 139
sequence space (6.1.1) 97
$s$-nuclear (11.2.1) 179
space, dual nuclear locally convex (4.1.6) 76
—, complete locally convex (0.5.7) 8
—, dual metric locally convex (0.7.5) 11
—, $(F)$- (0.7.4) 10
space, $(F')$- (0.7.5) 11
—, linear (0.3.1) 4
—, linearly semi-ordered (10.3.2) 175
—, locally convex (0.5.1) 6
—, metric (0.2.2) 3
—, metric locally convex (0.7.4) 10
—, $n$-dimensional linear (0.3.2) 4
—, normable (0.8.1) 11
—, normed (0.8.1) 11
—, nuclear locally convex (4.1.2) 70
—, precompact metric (0.2.5) 4
—, quasi-barrelled locally convex (0.7.1) 10
—, $\sigma$-quasi-barrelled locally convex (0.7.1) 10
—, reflexive locally convex (0.7.3) 10
—, separable topological (0.1.8) 2
—, semi-reflexive locally convex (0.7.2) 10
—, $s$-nuclear locally convex
—, topological (0.1.1) 1
—, universal nuclear locally convex (11) 177
spaces of analytic functions (6.4) 105
— of differentiable functions (6.2) 99
— of harmonic functions (6.3) 102
Spectral Decomposition Theorem (8.3.1) 129
step function (0.12.6) 17
strongly (0.6.8) 9
subset, absolutely convex (0.4.2) 5
—, absorbing (0.4.2) 5
—, bounded (0.5.5) 7
—, central (0.4.2) 5
—, closed (0.1.7) 2
—, compact (0.1.9) 3
—, dense (0.1.8) 2
—, equicontinuous (0.6.6) 9
—, essential (2.3.1) 38
—, measurable (0.12.4) 17
—, open (0.1.6) 2
—, precompact (0.2.5), (0.5.6) 4, 7
subspace, linear (0.3.3) 5
sum (1.1.1) 18
—, locally convex direct (5.2.2) 91
—, topological direct (5.2.2) 91
system, directed (0.1.5) 2

tensor product, algebraic (7.1.1) 108
—, locally convex (7.1.2) 108
topology (0.1.1) 1
—, coarser (0.1.3) 2
—, finer (0.1.3) 2

topology, induced (0.1.4) 2
—, natural (0.6.9) 10
—, strong (0.6.8) 9
—, weak (0.6.2) 8
type of a power series space (6.1.5) 99
type $l^p$ (8.2.1) 125
type $s$ (8.5.1) 138

unit ball, closed (0.8.1) 11

weakly (0.6.2) 8

zero neighborhood (0.5.1) 6

# Table of Symbols

| | | | | | |
|---|---|---|---|---|---|
| $\mathcal{A}(E, F)$ | 0.10.5 | $\boldsymbol{\varepsilon}_{(U, V)}(z)$ | 7.1.2 | $\mathcal{N}(E, F)$ | 3.1.2 |
| $\mathcal{A}_r(E, F)$ | 8.1.1 | $\hat{f}$ | 0.12.5 | $\boldsymbol{\nu}(T)$ | 3.1.1 |
| $\mathcal{A}(G)$ | 6.4.1 | $f_A$ | 0.12.6 | $\boldsymbol{\nu}^F(T)$ | 3.2.2 |
| $\alpha_r(T)$ | 8.1.1 | $\mathfrak{F}(I)$ | 1.1.1 | $\Omega_I$ | 1.5.7 |
| $\alpha_r^F(T)$ | 8.1.3 | $\mathfrak{F}_r(I)$ | 8.1.5 | $p(x)$ | 0.4.1 |
| $\mathcal{B}(E, F)$ | 7.4.1 | $\langle \varphi, \mu \rangle$ | 0.12.4 | $p[x(U)]$ | 0.11.1 |
| $\mathfrak{B}(E)$ | 0.5.5 | $\int_M \varphi(x)\, d\mu$ | 0.12.4 | $p_A(x)$ | 0.4.3 |
| $\mathfrak{B}_{\mathfrak{F}}(E)$ | 0.5.5 | $\Phi_I$ | 4.3.4 | $p_{(A, U)}(T)$ | 0.10.4 |
| $\boldsymbol{\beta}(T)$ | 0.10.4 | $\Phi_{\mathfrak{B}}(E)$ | 9.7.7 | $p'_B(a)$ | 0.6.8 |
| $\boldsymbol{c}_I$ | 1.1.6 | $\Phi_{\mathfrak{U}}(E)$ | 9.7.2 | $p'_U(a)$ | 0.8.1 |
| $\mathcal{C}(M)$ | 0.12.1 | $\mathcal{H}(G)$ | 6.3.2 | $P$ | 6.1.1 |
| $\mathcal{C}^+(M)$ | 0.12.3 | $\mathfrak{K}(E)$ | 0.5.6 | $\mathcal{P}(E, F)$ | 2.2.3 |
| $\mathcal{D}$ | 6.2.4 | $\boldsymbol{l}_I^1$ | 1.1.6 | $\mathcal{P}_0(E, F)$ | 3.2.8 |
| $\delta_r(B)$ | 9.1.1 | $\boldsymbol{l}_I^2$ | 1.1.7 | $\mathcal{P}_{\mathfrak{F}}(E)$ | 0.5.2 |
| $\delta_r(V, U)$ | 9.2.1 | $\boldsymbol{l}_I^1[E]$ | 1.2.2 | $\boldsymbol{\pi}(T)$ | 2.2.2 |
| $\delta_x$ | 2.3.1 | $\boldsymbol{l}_I^1(E)$ | 1.3.2 | $\boldsymbol{\pi}_0(T)$ | 3.2.3 |
| $\Delta_{\mathfrak{B}}(E)$ | 9.2.7 | $\boldsymbol{l}_I^1\{E\}$ | 1.4.2 | $\boldsymbol{\pi}_U[x_i, I]$ | 1.4.2 |
| $\Delta_{\mathfrak{U}}(E)$ | 9.2.2 | $\boldsymbol{l}_I^1\langle E \rangle$ | 1.5.2 | $\boldsymbol{\pi}_{(U, V)}(z)$ | 7.1.2 |
| $\Delta_I$ | 2.3.2 | $\boldsymbol{l}_I^1[E] = \boldsymbol{l}_I^1\{E\}$ | 4.2.1 | $\prod_I E_i$ | 5.2.1 |
| $E'$ | 0.6.1 | $\boldsymbol{l}_I^1[E] \equiv \boldsymbol{l}_I^1\{E\}$ | 4.2.1 | $R$ | 6.1.1 |
| $E'_b$ | 0.6.8 | $\boldsymbol{l}_I^1(E) = \boldsymbol{l}_I^1\{E\}$ | 4.2.1 | $\mathfrak{R}(T)$ | 0.10.1 |
| $E'_k$ | 0.6.7 | $\boldsymbol{l}_I^1(E) \equiv \boldsymbol{l}_I^1\{E\}$ | 4.2.1 | $\mathfrak{R}_\mu(M)$ | 0.12.4 |
| $E'_s$ | 0.6.2 | $\boldsymbol{l}^p(E, F)$ | 8.2.1 | $\varrho(V, U)$ | 9.8.1 |
| $E''$ | 0.6.9 | $\mathcal{L}(E, F)$ | 0.10.3 | $\boldsymbol{\varrho}_p(T)$ | 8.2.3 |
| $E''_n$ | 0.6.9 | $\mathcal{L}_b(E, F)$ | 0.10.4 | $\mathbf{s}(E, F)$ | 8.5.1 |
| $\tilde{E}$ | 0.5.7 | $\mathfrak{L}_\mu^2(M)$ | 0.12.6 | $\mathcal{S}$ | 6.2.5 |
| $\hat{E}$ | 0.5.7 | $\boldsymbol{\lambda}_\infty[\alpha_i, I]$ | 1.1.6 | $\mathcal{S}(E, F)$ | 2.5.2 |
| $E(A)$ | 0.11.2 | $\boldsymbol{\lambda}_1[\xi_i, I]$ | 1.1.6 | $\boldsymbol{\sigma}(T)$ | 2.5.1 |
| $E(A, B)$ | 0.11.5 | $\boldsymbol{\lambda}_2[\xi_i, I]$ | 1.1.7 | $\Sigma$ | 6.1.6 |
| $E(A, U)$ | 0.11.5 | $\boldsymbol{\lambda}_2(\varphi)$ | 0.12.5 | $\sum_I E_i$ | 5.2.2 |
| $E(U)$ | 0.11.1 | $\lambda_r(T)$ | 8.3.2 | $T'$ | 0.10.2 |
| $E(V, U)$ | 0.11.5 | $\Lambda(P)$ | 6.1.1 | $T^*$ | 0.12.2 |
| $E'(A^0)$ | 0.11.4 | $\boldsymbol{m}_I$ | 1.1.6 | $T_I[x_i, I]$ | 2.1.2 |
| $E'(U^0)$ | 0.11.3 | $M^0$ | 0.6.4 | $U_\varepsilon(x)$ | 0.2.2 |
| $E/F$ | 0.3.4 | $M^{00}$ | 0.6.4 | $\overline{U(F)}$ | 0.5.4 |
| $E \otimes F$ | 7.1.1 | $M_\varepsilon(B)$ | 9.6.1 | $\mathfrak{U}(E)$ | 0.5.1 |
| $E \otimes_\varepsilon F$ | 7.1.2 | $M_\varepsilon(V, U)$ | 9.7.1 | $\mathfrak{U}_{\mathfrak{F}}(E)$ | 0.5.1 |
| $E \otimes_\pi F$ | 7.1.2 | $\lvert\mu\rvert$ | 0.12.3 | $\mathfrak{U}_{\mathfrak{F}}(x)$ | 0.1.1 |
| $\mathcal{E}$ | 6.2.2 | $\mu(M)$ | 0.12.2 | $\langle x, a \rangle$ | 0.3.5 |
| $\mathcal{E}(\Delta)$ | 6.2.1 | $N$ | 1.5.5 | $x(F)$ | 0.3.4 |
| $\mathcal{E}_0(\Delta)$ | 6.2.3 | $N(U)$ | 0.11.1 | $x(U)$ | 0.11.1 |
| $\varepsilon_U[x_i, I]$ | 1.2.3 | $N < M$ | 0.11.5 | $x_i(\mathfrak{i})$ | 1.3.1 |

# Ergebnisse der Mathematik und ihrer Grenzgebiete

1. Bachmann: Transfinite Zahlen.
2. Miranda: Partial Differential Equations of Elliptic Type.
4. Samuel: Méthodes d'Algèbre Abstraite en Géométrie Algébrique.
5. Dieudonné: La Géométrie des Groupes Classiques.
7. Ostmann: Additive Zahlentheorie. 1. Teil: Allgemeine Untersuchungen.
8. Wittich: Neuere Untersuchungen über eindeutige analytische Funktionen.
11. Ostmann: Additive Zahlentheorie. 2. Teil: Spezielle Zahlenmengen.
13. Segre: Some Properties of Differentiable Varieties and Transformations.
15. Zeller/Beckmann: Theorie der Limitierungsverfahren.
16. Cesari: Asymptotic Behavior and Stability Problems in Ordinary Differential Equations.
17. Severi: Il teorema di Riemann-Roch per curve-superficie e varietà questioni collegate.
18. Jenkins: Univalent Functions and Conformal Mapping.
19. Boas/Buck: Polynomial Expansions of Analytic Functions.
20. Bruck: A Survey of Binary Systems.
23. Bergmann: Integral Operators in the Theory of Linear Partial Differential Equations.
25. Sikorski: Boolean Algebras.
26. Künzi: Quasikonforme Abbildungen.
27. Schatten: Norm Ideals of Completely Continuous Operators.
28. Noshiro: Cluster Sets.
29. Jacobs: Neuere Methoden und Ergebnisse der Ergodentheorie.
30. Beckenbach/Bellmann: Inequalities.
31. Wolfowitz: Coding Theorems of Information Theory.
32. Constantinescu/Cornea: Ideale Ränder Riemannscher Flächen.
33. Conner/Floyd: Differentiable Periodic Maps.
34. Mumford: Geometric Invariant Theory.
35. Gabriel/Zisman: Calculus of Fractions and Homotopy Theory.
36. Putnam: Commutation Properties of Hilbert Space Operators and Related Topics.
37. Neumann: Varieties of Groups.
38. Boas: Integrability Theorems for Trigonometric Transforms.
39. Sz.-Nagy: Spektraldarstellung linearer Transformationen des Hilbertschen Raumes.
40. Seligman: Modular Lie Algebras.
41. Deuring: Algebren.
42. Schütte: Vollständige Systeme modaler und intuitionistischer Logik.
43. Smullyan: First-Order Logic.
44. Dembowski: Finite Geometries.
45. Linnik: Ergodic Properties of Algebraic Fields.
46. Krull: Idealtheorie.
47. Nachbin: Topology on Spaces of Holomorphic Mappings.
48. A. Ionescu Tulcea/C. Ionescu Tulcea: Topics in the Theory of Lifting.
49. Hayes/Pauc: Derivation and Martingales.
50. Kahane: Séries de Fourier Absolument Convergents.
51. Behnke/Thullen: Theorie der Funktionen mehrerer komplexer Veränderlichen.
52. Wilf: Finite Sections of Some Classical Inequalities.

53. Ramis: Sous-ensembles analytiques d'une variété banachique complexe.
54. Busemann: Recent Synthetic Differential Geometry.
55. Walter: Differential and Integral Inequalities.
56. Monna: Analyse non-archimédienne.
57. Alfsen: Compact Convex Sets and Boundary Integrals.
58. Greco/Salmon: Topics in $\mathfrak{m}$-Adic Topologies.
59. López de Medrano: Involutions on Manifolds.
60. Sakai: C*-Algebras and W*-Algebras.
61. Zariski: Algebraic Surfaces.
62. Robinson: Finiteness Conditions and Generalized Soluble Groups, Part 1.
63. Robinson: Finiteness Conditions and Generalized Soluble Groups, Part 2.
64. Hakim: Topos annelés et schémas relatifs.
65. Browder: Surgery on Simply-Connected Manifolds.
66. Pietsch: Nuclear Locally Convex Spaces.
67. Dellacherie: Capacités et processus stochastiques.
68. Raghunathan: Discrete Subgroups of Lie Groups. In preparation.

202 · 100/560/72